Wave Factorization of Elliptic Symbols: Theory and Applications

Wave Factorization of Elliptic Symbols: Theory and Applications

Introduction to the Theory of Boundary Value Problems in Non-Smooth Domains

by

Vladimir B. Vasil'ev
Department of Mathematical Analysis,
Novgorod State University,
Novgorod, Russia

KLUWER ACADEMIC PUBLISHERS
DORDRECHT / BOSTON / LONDON

A C.I.P. Catalogue record for this book is available from the Library of Congress.

ISBN 978-90-481-5545-3

Published by Kluwer Academic Publishers,
P.O. Box 17, 3300 AA Dordrecht, The Netherlands.

Sold and distributed in the North, Central and South America
by Kluwer Academic Publishers,
101 Philip Drive, Norwell, MA 02061, U.S.A.

In all other countries, sold and distributed
by Kluwer Academic Publishers Group,
P.O. Box 322, 3300 AH Dordrecht, The Netherlands.

Printed on acid-free paper

Table of Contents

Preface

To summarize briefly, this book is devoted to an exposition of the foundations of pseudodifferential equations theory in non-smooth domains. The elements of such a theory already exist in the literature and can be found in such papers and monographs as [90,95,96,109,115,131,132,134,135,136,146, 163,165,169,170,182,184,214-218]. In this book, we will employ a theory that is based on quite different principles than those used previously. However, precisely one of the standard principles is left without change, the "freezing of coefficients" principle.

The first main difference in our exposition begins at the point when the "model problem" appears. Such a model problem for differential equations and differential boundary conditions was first studied in a fundamental paper of V.A.Kondrat'ev [134]. Here also the second main difference appears, in that we consider an **already given** boundary value problem. In some transformations this boundary value problem was reduced to a boundary value problem with a parameter λ in a domain with smooth boundary, followed by application of the earlier results of M.S. Agranovich and M.I. Vishik. In this context some operator-function $R(\lambda)$ appears, and its poles prevent invertibility; if for differential operators the function is a polynomial on λ, then for pseudodifferential operators this dependence on λ cannot be defined.

Ongoing investigations of different model problems are being carried out with approximately this plan, both for differential and pseudodifferential boundary value problems. "Non-smoothness" has, in the same way, been reduced to "domain with smooth boundary" for which its points play a parametric role (B.A. Plamenevsky [196], B.-W. Schulze [214-218]).

For composite singularities of top of polyhedral angle type, the given boundary value problem was also reduced to a boundary value problem in a space of lower dimension (V.G. Maz'ya, B.A. Plamenevsky [163]) i.e., on a strate (wedge); then the unique solvability of such an auxiliary problem was postulated. A.I. Komech [132] proved the validity of this postulate in only some cases .

I suggest considering the model problem in a different manner, as a multidimensional analogue of the M.I. Vishik-G.I. Eskin approach [92,93,281,282], which was applied to the study of pseudodifferential equations and boundary value problems in a domain with smooth boundary. A model pseudodifferential equation in this case is an equation in a half-space which, with the help of factorization (Wiener-Hopf method), can be solved explicitly (i.e., with either a general solution or solvability conditions written on the right-hand side of the equation). The last alternative is determined by the sign of the so-called index of factorization (for more details, see chapter 4). The form of a general solution, or the character of solvability conditions, prompted the statement of a boundary value problem or restatement of the problem with additional potentials. So the model equation, and more precisely the character of its solvability, was described with maximal sharpness, and it determined an exhaustive answer to the question about solvability of boundary value problems for a pseudodifferential equation in a domain with smooth boundary.

My method also involves a type of factorization called "wave factorization." Its essential difference from the above is in its multidimensionality. In the simplest model domain, such as an infinite angle in a plane, this factorization is connected with analyticity in the domain of **two-dimensional complex space.** This complication, however, resulted in a striking analogue with the theory of M.I. Vishik-G.I. Eskin,specifically the same solvability picture for a model pseudodifferential equation in an angle that is dependent on the index of wave factorization. The description of this factorization and its application to the study of model pseudodifferential equations (which often appear in applications), had no explicit analytical expression until now [7,8,23,24,166-168]). The subject matter of this book is therefore the correct statement of boundary value problems, consideration of problems with additional potentials, a series of "standard" examples, and construction for them of explicit solutions. It is presented in the hope that wave factorization will be helpful to others involved in such studies and will have a long and useful life. The core of the material in the book is accessible to mathematics graduate students; material not usually offered in university courses is described in a number of special chapters.

A brief word on the structure of the book. The first four chapters include the material which one needs to read the following chapters. The reader must pay special attention to Chapter 2 in which selected facts from multidimensional complex analysis are collected (this material stands far apart from the other "analytical" material used). Chapter 5 is devoted to the basic concept of wave factorization. Chapters 6,7 are devoted to two applied problems from diffraction and elasticity where the wave factorization method permits us to find the solution in a precise form.

In chapter 8, with the help of wave factorization, the solvability picture of model pseudodifferential equation in a plane angle is obtained. Because a general solution always includes arbitrary functions, one needs additional conditions for its unique determination. These conditions are suggested in chapter 9 as traces of some pseudodifferential equations on angle sides, and then such a boundary value problem is reduced to an equivalent system of integral equations. In the general case, solution of this system is very hard, but in some interesting situations (when symbols are homogeneous and boundary conditions are specific) it is possible with the help of the Mellin transform (chapter 10). In chapter 11 the "classic case" is considered, the one that is Laplacian with Dirichlet or Neumann conditions on angle sides. In this case, with the help of tables, all Mellin transforms of kernels of integral equations can be calculated.

In chapter 12 the problem is considered with additional potentials (co-boundary operators); its statement requires the negative index of wave factorization.

Originally I intended to include in the book an additional chapter related to parabolic equations. But as of now this material is still in quite raw form, and one of my students is occupied with giving it an acceptable form. I have included three appendices devoted to results on multidimensional singular integral operators in a non-smooth domain up to the index theorem. These results are interesting enough in their own right and have a natural relation to the concept of wave factorization.

Appendix 1 contains the "philosophy" of multivariable variant of the Wiener-Hopf method. The main result of Appendix 2 is the theorem on vanishing of the index of an elliptic singular integral operator. In Appendix 3, properties of the Mellin transform are collected for the reader's convenience.

With regard to the references listed in the book, none of the papers mentioned (except for my own) concerns the book directly. But all of these papers, in the aggregate, have influenced what has been said in the book.It is clear that most of these papers are related to pseudodifferential and singular integral operators (or corresponding equations and boundary value problems) on manifolds with singularities.

Finally, I would like to thank Dr. Paul Roos, his assistant Ms Angelique Hempel, language editors Mr Edwin Beschler and Mr George Novosyolov, editorial assistant Ms Anneke Pot and Kluwer Academic Publishers for their personal support that led to the writing of this book. Without their help, it would have appeared much later.

I am also grateful to mathematician Vladimir G. Utkin for his hard work involved in type-setting of the book.

Vladimir B. Vasil'ev Novgorod, 1998

1. Distributions and Their Fourier transforms

1.1. TEMPERED DISTRIBUTIONS

Distributions are linear continuous functionals over the space of so-called basic functions. Space of basic functions we choose as the Schwartz space $S(\mathbb{R}^m)$ ($\mathbb{R}^m$ is m - dimensional Euclidean space) of infinitely differentiable on $\mathbb{R}^m$ functions decreasing under $|x| \to \infty$ more rapidly than any power of $|x|^{-1}$, $x = (x_1, \dots, x_m)$, $|x| = \sqrt{x_1^2 + \dots + x_m^2}$. We determine the counting number of norms in $S(\mathbb{R}^m)$ by the formula

$$||\varphi||_p = \sup_{|\alpha| \leq p} (1 + |x|^2)^{p/2} |\mathcal{D}^\alpha \varphi(x)|, \quad \varphi \in S(\mathbb{R}^m), \; p = 0, 1, \dots , \tag{1.1.1}$$

where $\mathcal{D}^\alpha \varphi = \frac{\partial^{|\alpha|} \varphi}{\partial x_1^{\alpha_1} \dots \partial x_m^{\alpha_m}}$, α is multi index, $|\alpha| = \alpha_1 + \dots + \alpha_m$; with the help of these norms we define the convergence concept in $S(\mathbb{R}^m)$. Namely we say the sequence $\varphi_1, \dots, \varphi_k, \dots$ of functions from $S(\mathbb{R}^m)$ converges to function $\varphi \in S(R^m)$ iff $||\varphi_k - \varphi||_p \to 0$, $k \to \infty$, for all $p = 0, 1, \dots$ The last statement, by virtue of (1.1.1) is equivalent to saying that $x^\alpha \mathcal{D}^\beta \varphi_k(x)$ uniformly tends to zero under $k \to \infty$ for arbitrary multiindex α, β, $x^\alpha \equiv x_1^{\alpha_1} \dots x_m^{\alpha_m}$.

After in the space $S(\mathbb{R}^m)$ convergence is defined (and hence topology is introduced) one can give the following.

Definition 1.1.1. *An arbitrary linear continuous functional on the Schwartz space $S(\mathbb{R}^m)$ is called a tempered distribution. The space of tempered distributions is denoted by $S'(\mathbb{R}^m)$.*

Value of $f \in S'(\mathbb{R}^m)$ on $\varphi \in S(\mathbb{R}^m)$ are usually denoted by (f, φ).

We give two of the most widely occuring examples of tempered distributons.

Example 1.1.1. Let $f(x)$ be locally integrable, and for some $n \geq 0$

$$\int_{\mathbb{R}^m} |f(x)|(1 + |x|)^{-n} dx < +\infty.$$

Then $f(x)$ defines $f \in S'(\mathbb{R}^m)$ by formula

$$(f, \varphi) = \int_{\mathbb{R}^m} f(x)\varphi(x) dx, \quad \forall \varphi \in S(\mathbb{R}^m). \tag{1.1.2}$$

Linearity of the functional (1.1.2) is obvious, but its continuity follows from inequality

$$\begin{aligned} |(f, \varphi)| &= \left| \int_{\mathbb{R}^m} |f(x)|(1 + |x|)^{-n} \varphi(x) dx \right| \leq \\ &\leq \sup (1 + |x|)^n |\varphi(x)| \int_{\mathbb{R}^m} |f(x)|(1 + |x|)^{-n} dx = c||\varphi||_n \end{aligned} \tag{1.1.3}$$

(here and below the "c" will denote some positive constant).

Definition 1.1.2. *The tempered distributions defined by formula (1.1.2), are called regular.*

Example 1.1.2 (Dirac δ - function). Let us define the linear continuous functional $\delta \in S'(\mathbb{R}^m)$ on $S(\mathbb{R}^m)$ by the formula

$$(\delta, \varphi) = \varphi(0).$$

Below we will verify that Dirac δ - functions can't be represented by formula (1.1.2), i.e. there is no locally integrable function $f(x)$ such that

$$\int_{\mathbb{R}^m} f(x)\varphi(x)dx = \varphi(0), \quad \forall \varphi \in S(\mathbb{R}^m).$$

The Dirac δ - function is a clear example of a singular tempered distribution.

1.2. OPERATIONS WITH DISTRIBUTIONS

We give a far from complete list of operations with tempered distributions, selecting only those that seem the most important for our purposes.

Definition 1.2.1. *The derivative $\mathcal{D}^\alpha f$ of a tempered distribution $f \in S'(\mathbb{R}^m)$ is a tempered distribution satisfying the equlity*

$$(\mathcal{D}^\alpha f, \varphi) = (-1)^{|\alpha|}(f, \mathcal{D}^\alpha \varphi), \forall \varphi \in S(\mathbb{R}^m). \tag{1.2.1}$$

Obviously the formula (1.2.1) defines a linear continuous functional on $S(\mathbb{R}^m)$, i.e., the inequality of (1.1.3) type holds.

Defininition 1.2.1 is motivated by the following elementary fact.

If $f(x) \in S(\mathbb{R}^m)$ then

$$\left(\frac{\partial f}{\partial x_i}, \varphi\right) = \int_{\mathbb{R}^m} \frac{\partial f}{\partial x_i}(x)\varphi(x)dx,$$

and after integration by parts one obtains

$$\left(\frac{\partial f}{\partial x_i}, \varphi\right) = -\int_{\mathbb{R}^m} f(x)\frac{\partial f}{\partial x_i}(x)dx = -\left(f, \frac{\partial f}{\partial x_i}\right), \quad i = 1, 2, \ldots, m.$$

Here the essential point is that an arbitrary tempered distribution is infinitely differentiable.

Let $f \in S'(\mathbb{R}^m)$, $a(x) \in C^\infty(\mathbb{R}^m)$ (the class of infinitely differentiable on $\mathbb{R}^m$ functions). Then the formula

$$(af, \varphi) = (f, \overline{a}\varphi), \quad \forall \varphi \in S(\mathbb{R}^m), \tag{1.2.2}$$

where $\overline{a}(x)$ is the complex conjugate function to $a(x)$, defines a tempered distribution if

$$|\mathcal{D}^r a(x)| \le c_r(1+|x|)^{k_r}, \quad 0 \le |r| < \infty.$$

The last estimate is needed for proof of continuity for functional (1.2.2).

If $a \in \mathbb{R}^m$ is an arbitrary vector, then by a shift of the tempered distribution $f \in S'(\mathbb{R}^m)$ one finds the tempered distribution f^a which is defined by formula

$$(f^a, \varphi) = (f, \varphi(x+a)). \tag{1.2.3}$$

The definition (1.2.3) is also motivated by its validity for "good" f and regular representation (1.1.2).

More generally, if L is an $m \times m$ - matrix and $\det L \neq 0$, one can define the tempered distribution $f^{L,b}$ by

$$(f^{L,b}, \varphi) = (f, \varphi(Lx+b)), \quad \forall \varphi \in S(\mathbb{R}^m), \tag{1.2.3'}$$

i.e., (1.2.3) is a linear change of variables formula for distributions.

Let us note that all given operations (differentiation, multiplication on infinitely differentiable functions, change of variables) are linear and continuous from $S'(\mathbb{R}^m)$ to itself.

Let $x = (x', x_m)$, $x' = (x_1, \dots, x_{m-1}) \in \mathbb{R}^{m-1}$, and let f_1, f_2 be regular distributions, $f_1 \in S(\mathbb{R}^{m-1})$, $f_2 \in S(\mathbb{R})$.

Definition 1.2.2. *The direct product $f_1 \otimes f_2$ is the tempered distribution defined by formula*

$$(f_1 \otimes f_2, \varphi) = \int_{\mathbb{R}^{m-1}} \int_{-\infty}^{+\infty} f_1(x') f_2(x_m) \overline{\varphi(x', x_m)} dx' dx_m. \tag{1.2.4}$$

Obviously (1.2.4) correctly defines the regular tempered distribution $f_1 \otimes f_2 \in S'(\mathbb{R}^m)$.

Simple arguments permit us to enlarge the direct product construction $\otimes$ on arbitrary tempered distributions $f_1 \in S'(\mathbb{R}^m)$, $f_2 \in S'(\mathbb{R}^k)$ obtaining as a result the tempered distribution $f_1 \otimes f_2 \in S'(\mathbb{R}^{m+k})$.

The convolution operation for tempered distributions is not always defined. But if $f \in S(\mathbb{R}^m)$ then one can give

Definition 1.2.3. *The convolution of a distribution $f \in S'(\mathbb{R}^m)$ and a basic function $\varphi \in S(\mathbb{R}^m)$ is the function*

$$f * \varphi = \left(f(y), \overline{\varphi(x-y)} \right). \tag{1.2.5}$$

Some explanations are needed here. In formula (1.2.5) the variable x is fixed, and the word "function" rather than "distribution" is used deliberately. Actually (1.2.5) defines the infinitely differentiable on $\mathbb{R}^m$ function which grows at infinity (with its all derivatives) no more rapidly than a power of $(1+|x|)$.

1.3. SUPPORT OF DISTRIBUTIONS

Recall that the support of a continuous function $\varphi(x)$ on $\mathbb{R}^m$ is the closure of points in which $\varphi(x) \neq 0$. Although distributions are linear functionals, nevertheless one can define vanishing of a functional on an open set in the following sense.

Let $U \subset \mathbb{R}^m$ be an open set. Denote $S(U)$ the space of infinitely differentiable compactly supported functions for which their supports belong to U; precisely, $S(U)$ is the closure of these functions in the $S(\mathbb{R}^m)$ - topology.

Definition 1.3.1. *A distribution defined in a domain U is called a linear continuous functional on the space $S(U)$. The space of distributions defined in domain U is denoted $S'(U)$.*

Let $U_1 \subset \mathbb{R}^m$, $U_2 \subset \mathbb{R}^m$ be two open sets, and $U_1 \subset U_2$. It is not difficult to guess that $S'(U_2) \subset S'(U_1)$, and convergence in $S'(U_2)$ implies convergence in $S'(U_1)$. This leads to an important conclusion: For every distribution $f \in S'(U)$ there is an unique restriction on arbitrary open sets $U' \subset U$ such that $f \in S'(U')$.

Definition 1.3.2. *Distribution $f \in S'(U)$ vanishes in open set $U' \subset U$ iff its restriction on U' is zero functional from $S'(U')$ i.e., $(f,\varphi) = 0$, $\forall \varphi \in S(U')$. Notation: $f(x) = 0$, $x \in U'$.*

With the help of simple arguments and the Heine-Borel lemma on finite covering one can prove

Proposition 1.3.1. If a distribution from $S'(U)$ vanishes in some neighbourhood of every point of the open set U, then it vanishes in the whole set U.

Let us denote by U_f the union of all neighbourhoods of points in which $f = 0$. U_f is an open set, by Proposition 1.3.1 $f = 0$ in U_f, and U is the extremal set in which f vanishes.

Definition 1.3.3. *The complement of U_f up to U is called the support of distribution f.*

The support of f is denoted by $\operatorname{supp} f$, it is a closed set in U.

In the case of a regular distribution, the question of when it vanishes in domain U has a simple answer.

Proposition 1.3.2. *A function f that is locally integrable in U vanishes almost everywhere in U in and only if a regular distribution f generated by it vanishes in U.*

Proposition 1.3.2 implies taht there is a one-to-one correspondence between locally integrable in U functions and regular distributions in U.

To finish this part using the above definitions and propositions, let us return to example 1.1.2 and verify that the Dirac δ - function is a singular distribution.

Suppose the other, i.e., there is alocally integrable on $\mathbb{R}^m$ function f such that

$$\int_{\mathbb{R}^m} f(x)\varphi(x)dx = (\delta,\varphi) = \varphi(0), \quad \forall \varphi \in S(\mathbb{R}^m). \tag{1.3.1}$$

As $|x|^2\varphi(x) \in S(\mathbb{R}^m)$ then (1.3.1) implies

$$\int\limits_{\mathbb{R}^m} f(x)|x|^2\varphi(x)dx = |x|^2\varphi(x)\big|_{x=0} = 0 = \left(|x|^2 f, \varphi\right),$$

and it holds for every $\varphi \in S(\mathbb{R}^m)$. Thus the locally integrable function $|x|^2 f(x)$ is equal to zero in the distribution sense, and according to Proposition 1.3.2, $|x|f(x) = 0$ almost everywhere $\Rightarrow f(x) = 0$ almost everywhere in $\mathbb{R}^m$. The last contradicts equality (1.3.1).

Finishing here with the δ - function we note that $\sup \delta = \{0\}$, and the following proposition can be stated.

Proposition 1.3.3. *If the support of distribution $f \in S'(\mathbb{R}^m)$ consists of a unique point $x = 0$, then it is uniquely represented in the form*

$$f(x) = \sum_{|\alpha| \leq N} c_\alpha D^\alpha \delta(x),$$

where N is order of f, and c_α are some constants.

1.4. FOURIER TRANSFORM OF DISTRIBUTIONS

For basic functions from the space $S(\mathbb{R}^m)$, the classical Fourier transform is denoted:

$$(F\varphi)(\xi) \equiv \tilde{\varphi}(\xi) = \int\limits_{\mathbb{R}^m} \varphi(x)e^{ix\cdot\xi}dx, \quad \varphi \in S(\mathbb{R}^m), \tag{1.4.1}$$

$x \cdot \xi = x_1\xi_1 + \cdots + x_m\xi_m$.

So the defined function $\tilde{\varphi}(\xi)$ is bounded and continuous on $\mathbb{R}^m$. It is possible to differentiate under the integral sign, and from this follows

$$\mathcal{D}^\alpha(F\varphi)(\xi) = \int\limits_{\mathbb{R}^m} (ix)^\alpha \varphi(x)e^{ix\cdot\xi}dx = F\left((ix)^\alpha\varphi\right)(\xi). \tag{1.4.2}$$

Analogously,

$$F(\mathcal{D}^\alpha\varphi)(\xi) = \int\limits_{\mathbb{R}^m} \mathcal{D}^\alpha\varphi(x)e^{ix\cdot\xi}dx = (-i\xi)^\alpha(F\varphi)(\xi). \tag{1.4.3}$$

The inverse Fourier transform has the form

$$\left(F^{-1}\psi\right)(x) = \frac{1}{(2\pi)^m}\int\limits_{\mathbb{R}^m} \psi(\xi)e^{-ix\cdot\xi}d\xi. \tag{1.4.4}$$

Proposition 1.4.1. *The Fourier transform operator is one-to-one and continuously maps the space $S(\mathbb{R}^m)$ on itself.*

We note some other properties of the Fourier transform.
a) $F(\varphi * \psi) = (F\varphi)\cdot(F\psi)$ (Fourier transform of convolution);
b) $F(\varphi(x-a)) = e^{ia\cdot\xi}(F\varphi)(\xi)$ (Fourier transform of shift);
c) Parseval equality

$$\int_{\mathbb{R}^m} \varphi\overline{\psi(x)}dx = \frac{1}{(2\pi)^m}\int_{\mathbb{R}^m} \tilde{\varphi}(\xi)\overline{\tilde{\psi}(\xi)}d\xi,$$

and taking $\varphi = \psi$,

$$\int_{\mathbb{R}^m} |\varphi(x)|^2 dx = \frac{1}{(2\pi)^m}\int_{\mathbb{R}^m} |\tilde{\varphi}(\xi)|^2 d\xi. \tag{1.4.5}$$

Very important for further investigation is the following property of the Fourier transform.

Let $\mathbb{R}^m_+$ be the half-space $x_m > 0$. Denote $S^+(\mathbb{R}^m)$ as the subspace in $S(\mathbb{R}^m)$ consisting of functions $\varphi(x', x_m) \in S(\mathbb{R}^m)$ vanishing under $x_m < 0$.

Proposition 1.4.2. *The Fourier-image $\tilde{S}^+(\mathbb{R}^m)$ of subspace $S^+(\mathbb{R}^m)$ consists of functions $\tilde{\varphi}(\xi', \xi_m)$ admitting analytical continuation on ξ_m into the complex half-plane $\xi_m + i\tau, \tau > 0$, which are infinitely differentiable under $(\xi', \xi_m) \in \mathbb{R}^m$, $\tau \geq 0$, and satisfying the following estimates*

$$(1 + |\xi'| + |\xi_m| + \tau)^m \left|\mathcal{D}^{\alpha'}_{\xi'}\mathcal{D}^{\alpha_m}_{\xi_m}\tilde{\varphi}(\xi', \xi_m)\right| \leq c_{m,|\alpha|},$$
$$0 \leq m < +\infty, \quad \alpha = (\alpha', \alpha_m), \quad 0 \leq |\alpha| < +\infty.$$

An analogous proposition is valid for $S^-(\mathbb{R}^m)$.

For a pair of functions $\varphi, \psi \in S(\mathbb{R}^m)$ it is easy to verify by the Fubini theorem the following equality:

$$\int_{\mathbb{R}^m} \tilde{\varphi}(\xi)\psi(\xi)d\xi = \int_{\mathbb{R}^m} \varphi(x)\tilde{\psi}(x)dx,$$

which we take as our definition of the Fourier transform for a tempered distribution

Definition 1.4.1. *A Fourier transform of tempered distribution $f \in S'(\mathbb{R}^m)$ is tempered distribution $\tilde{f} \in S'(\mathbb{R}^m)$ that*

$$(\tilde{f}, \varphi) = (f, \tilde{\varphi}), \quad \forall\varphi \in S(\mathbb{R}^m). \tag{1.4.6}$$

Example 1.4.1. The Fourier transform for

$$f(x) = \sum_{|\alpha|\leq N} c_\alpha D_\alpha \delta(x)$$

is the polynomial

$$\tilde{f}(x) = \sum_{|\alpha|\leq N} c_\alpha \xi^\alpha.$$

So defined, the Fourier transform of tempered distributions, as we have anticipated, inherits many properties of the classical Fourier transform. Thus let us look at some nuances.

By formula (1.2.5) we defined convolution for basic functions and distributions.

Proposition 1.4.3. *Let* $f \in S'(\mathbb{R}^m)$, $\varphi \in S(\mathbb{R}^m)$. *Then*

$$F(f * \varphi) = \tilde{\varphi}(\xi) \cdot \tilde{f}. \tag{1.4.7}$$

The right-hand side of (1.4.7) is the distribution defined by formula (1.2.2).

By formula (1.2.4) we defined the direct product of tempered distributions.

Proposition 1.4.4. Let $f_1 \in S'(\mathbb{R}^{m-1})$, $f_2 \in S'(\mathbb{R})$. Then

$$F(f_1 \otimes f_2) = \tilde{f}_1 \otimes \tilde{f}_2. \tag{1.4.8}$$

2. Multidimensional complex analysis

2.1. CONES IN THE SPACE $\mathbb{R}^m$

Definition 2.1.1. *A* **cone** $C \subset \mathbb{R}^m$ *(with top at the origin) is point set which has property: if* $y \in C$ *then* $\lambda y \in C$ *for all* $\lambda > 0$.

The intersection of cone C *with sphere* $|x| = 1$ *is called its* **projection** *and is denoted by* pr. C. *The cone* C' *such that* pr. $\overline{C'} \subset$ pr. C *is called a* **compact** *subcone of cone* C : *notation* $C' \subset\subset C$.

For the cone C *let us introduce the* **conjugate cone**

$$\overset{*}{C} = \{x \in \mathbb{R}^m : \ x \cdot y \geq 0, \ y \in C\}.$$

If $\overset{*}{C} = \overline{C}$ *then the cone* C *is called* **self-conjugate**.

The **indicatrix** *of the cone* C *is the function*

$$\mu_C(x) = \sup_{y \in \text{pr.}\, C} (-x \cdot y).$$

The simplest example is a right circular cone with top at the origin

$$C_+^a = \{x \in \mathbb{R}^m : \ x_m > a|x'|, \ a > 0\}$$

which we will use below, and

$$\overset{*}{C}{}_+^a = \{x \in \mathbb{R}^m : \ ax_m > |x'|, \ a > 0\}.$$

If $a = 1$ the cone C_+^1 is self-conjugate. The cones C_+^1 and $-C_+^1$ are called light cones of future and past respectively.

There are some relations between functional characteristics of the cone and its geometric properties. Let us describe some of them.

Proposition 2.1.1. $\overline{C} = \overline{C}_1$ *iff* $\mu_C(x) = \mu_{C_1}(x)$.

Thus, an indicatrix "distinguishes" the cones.

Let us denote by $\mathrm{ch}\, C$ the **convex hull** of cone C, $\overset{*}{C} = \{x \in \mathbb{R}^m :\ \mu_C(x) > 0\}$, and denote the number

$$\rho_C = \sup_{x \in \overset{*}{C}} \frac{\mu_{\mathrm{ch}\, C}(x)}{\mu_C(x)}.$$

This is a characteristic of the non-convexity of cone C. In fact, the following holds.

Proposition 2.1.2. Cone C is convex iff $\rho_C = 1$.

Indeed if the cone C is convex, then $\mathrm{ch}\, C = C$, and $\rho_C = 1$. Let us verify the inverse, and note the obvious inequality $\mu_C(x) \leq \mu_{\mathrm{ch}\, C}(x)$, $\forall x$. It turns out to be

$$\mu_C(x) \leq \mu_{\mathrm{ch}\, C}(x), \quad x \in \overset{*}{C}. \tag{2.1.1}$$

By definition $\mu_{\mathrm{ch}\, C}(x) \leq 0, x \in \overset{*}{C}$, and then $x \cdot y \geq 0$ for all $y \in \mathrm{ch}\, C$. Since $x \cdot y$ is a linear function on y its extremal values on any set and its convex hull coincide (here the homogeneity is important but not linearity, and linearity implies homogeneity). Thus,

$$\mu_C(x) = -\inf_{y \in \mathrm{pr.}\, C} x \cdot y = -\inf_{y \in C \setminus B(0;1)} x \cdot y =$$

$$= -\inf_{y \in \mathrm{ch}\, [C \setminus B(0;1)]} x \cdot y \geq -\inf_{y \in \mathrm{ch}\, C \setminus B(0;1)} x \cdot y =$$

$$= -\inf_{y \in \mathrm{ch}\, C} x \cdot y = \mu_{\mathrm{ch}\, C}(x), \quad x \in \overset{*}{C},$$

where $B(0;1)$ is the unit ball with center at 0, and $\mathrm{ch}\, C \setminus B(0;1) \subset \mathrm{ch}\, [C \setminus B(0;1)]$.

So, (2.1.1) is proved. ■

This implies that $\mu_C(x) = \mu_{\mathrm{ch}\, C}(x)$ for all $x \in \mathbb{R}^m$, and according to Proposition 2.1.1, $\overline{C} = \overline{\mathrm{ch}\, C}$, which implies convexity of C.

Definition 2.1.2. *The cone C is called a sharp cone if there is a plane of support to $\overline{\mathrm{ch}\, C}$ which has with $\overline{\mathrm{ch}\, C}$ only one common point* 0.

Proposition 2.1.3. *The following assertions are equivalent:*

a) *cone C is sharp;*

b) *cone $\overline{C}$ doesn't contain a whole straight line;*

c) *for arbitrary $C' \subset\subset \mathrm{int}\, \overset{*}{C}$ (*int *denotes the interior) there is a number $\delta = \delta(C') > 0$ such that*

$$x \cdot y \geq \delta \cdot |x| \cdot |y|, \quad y \in C', \ x \in \overline{\mathrm{ch}\, C};$$

d) *for any $y \in \mathrm{pr.int}\, \overset{*}{C}$ the set*

$$C_y = \{x \in \mathbb{R}^m :\ 0 \leq x \cdot y \leq 1,\ x \in \overline{\mathrm{ch}\, C}\}$$

is bounded.

Example 2.1.1. Let us return to cone C_+^1 once again. Then $\overset{*}{C_+^1} = C_+^1$, $\rho_{C_+^1} = 1$,

$$\mu_{C_+^1} = \begin{cases} |x|, & x \in -C_+^1 \\ \frac{1}{\sqrt{2}}(|x'| - x_m), & x \notin -C_+^1 . \end{cases}$$

If we denote $C_-^1 = -C_+^1$ and we put $C = C_+^1 \cup C_-^1$, then $\operatorname{ch} C = \mathbb{R}^m$, $\overset{*}{C} = \{0\}$, $\rho_C = \sqrt{2}$, $\mu_{\operatorname{ch} C}(x) = |x|$,

$$\mu_C(x) = \begin{cases} |x|, & x \in C \\ \frac{1}{\sqrt{2}}(|x'| + x_m), & x \notin C . \end{cases}$$

2.2. RADIAL TUBE DOMAINS AND ANALYTICAL FUNCTIONS

By an m - dimensional complex space $\mathbb{C}^m$ we mean a point collection of type

$$z = x + iy = (z_1, z_2, \ldots, z_m) = (x_1 + iy_1, x_2 + iy_2, \ldots, x_m + iy_m),$$
$$z_j \in \mathbb{C}, \quad j = 1, 2, \ldots, m.$$

In other words $\mathbb{C}^m = \mathbb{R}^m + i\mathbb{R}^m$.

Definition 2.2.1. *The cone $T(C) = \mathbb{R}^m + iC$, where C is an open cone in $\mathbb{R}^m$, is called the tube cone. If the cone C is connected, then $T(C)$ is called the radial tube domain (over the cone C).*

Definition 2.2.2. *Function $f(z)$ is called analytical (holomorphic) at point $z^\circ \in \mathbb{C}^m$ if in some neighbourhood of this point there exist all partial derivatives of first order $\frac{\partial f}{\partial \bar{z}_j}$, $j = 1, 2, \ldots, m$, i.e., the Cauchy-Riemann conditions hold:*

$$\frac{\partial u}{\partial x_j} = \frac{\partial v}{\partial y_j}, \quad \frac{\partial u}{\partial y_j} = -\frac{\partial v}{\partial x_j}, \quad j = 1, 2, \ldots, m,$$
$$f = u + iv.$$

We are interested in functions that are analytical in radial tube domains over cones. Let us begin with some concrete functions that later will play the role of very important examples. To describe these functions we need one auxiliary assertion.

Lemma 2.2.1. *The point $z = (z_1, \ldots, z_m) \in \mathbb{C}^m$ belongs to $\overset{*}{C^a} \equiv \overset{*}{C_+^a} \cup \overset{*}{C_-^a}$ if and only if*

$$a^2(z_m - x_m)^2 - (z_{m-1} - x_{m-1})^2 - \ldots - (z_1 - x_1)^2 \neq \rho, \quad \rho \geq 0,$$

for all $x \in \mathbb{R}^m$.

Proof. For a brief notation let us denote

$$z^2 = a^2 z_m^2 - z_{m-1}^2 - \cdots - z_1^2, \quad z = (z', z_m), \ z' = (z_1, \ldots, z_{m-1}).$$

Let $z = x' + iy$. From condition

$$(x' - x + iy)^2 = (x' - x)^2 - y^2 + 2iy(x' - x) \neq \rho \geq 0, \quad \forall x \in \mathbb{R}^m, \qquad (2.2.1)$$

follows that $y^2 \neq 0$. If we have $y^2 < 0$ then choosing x such that

$$\begin{aligned} &x'_m = x_m \\ &y_1(x_1 - x'_1) + \cdots + y_{m-1}(x_{m-1} - x'_{m-1}) = 0, \\ &(x_1 - x'_1)^2 + \cdots + (x_{m-1} - x'_{m-1})^2 = -y^2, \end{aligned}$$

we obtain from (2.2.1) $(z - x')^2 = 0$, and which is impossible. So, $y^2 > 0$, i.e., $a^2 y_m^2 - y_{m-1}^2 - \cdots - y_1^2 > 0$, and it means $z \in \overset{*}{C}{}^a$.

Conversely, if $y^2 > 0$ then either

$$y_1(x_1 - x'_1) + \cdots + y_{m-1}(x_{m-1} - x'_{m-1}) = 0,$$

or under $y \cdot (x - x') = 0$,

$$(x' - x)^2 - y^2 = \left[\frac{\overline{y} \cdot (\overline{x'} - \overline{x})}{y_m}\right]^2 - |\overline{x} - \overline{x'}|^2 - y^2 \leq$$

$$\leq -y^2 \left[1 + \frac{|\overline{x} - \overline{x'}|^2}{y_m^2}\right] < 0$$

(here $\overline{x} = (x_1, \ldots, x_{m-1})$, $\overline{y} = (y_1, \ldots, y_{m-1})$), and (2.2.1) holds. ■

Corollary 2.2.1. *The function $(-z^2)^p$, where p is an arbitrary real number, is analytical and univalent in $\overset{*}{C}{}^a$.*

Indeed $(-z^2)^p$ is analytical as a composition of two analytical functions $\zeta = -z^2$ $(\zeta = \xi + i\eta,\ \xi, \eta \in \mathbb{R})$ and $\omega = \zeta^p$, $\overset{*}{C}{}^a \xrightarrow{\zeta} \mathbb{C} \setminus \{\xi \leq 0,\ \eta = 0\} \xrightarrow{\omega} \mathbb{C}$, and since the first doesn't take negative values, the second has no bifurcation points.

Definition 2.2.3. *Let a function $f(z)$ be analytical in the tube domain $T(C) = \mathbb{R}^m + iC$. A* **spectral function** *of function $f(z)$ is the distribution $g \in S'(\mathbb{R}^m)$ which has the properties:*

$$\begin{aligned} &\text{a) } g(\xi)e^{-y \cdot \xi} \in S'(\mathbb{R}^m), \quad \forall y \in C, \\ &\text{b) } f(z) = F\left(g(\xi)e^{-y \cdot \xi}\right)(x) = \int_{\mathbb{R}^m} g(\xi) e^{iz \cdot \xi} d\xi \end{aligned}$$

for all $z \in T(C)$, $z = x + iy$.

The function $f(z)$ is called the Fourier-Laplace transform of the spectral function $g(\xi)$. A support of spectral function g is called a spectra of function $f(z)$.

Proposition 2.2.1. *In order that the function $f(z)$, which is analytical in domain $T(C)$, is the Fourier-Laplace transform of some spectral function, it is* *sufficient for every compact set $K \subset C$ there are some numbers $M = M(K)$ and $m = m(K)$ such that*

$$|f(z)| \leq M(1+|x|)^m, \quad z \in \mathbb{R}^m + iK; \tag{2.2.2}$$

and it is necessary that $f(z)$ is an analytical function in the convex hull $\operatorname{ch} T(C) = T(\operatorname{ch} C)$ *and that* **all its derivatives together** *satisfy the estimate of (2.2.2) type.*

Corollary 2.2.2. *If $f(z)$ is analytical in $T(C)$ and satisfies the inequality (2.2.2), then $f(z)$ is analytical in $T(\operatorname{ch} C)$ and satisfies in $T(\operatorname{ch} C)$ the inequality of (2.2.2.) type with all its derivatives together.*

In the following assertion we have a criteria of that the given function g must be a spectral function of function $f(z)$ which is analytical in domain $T(C)$.

Proposition 2.2.2. *In order for the distribution $g(\xi) \in S'(\mathbb{R}^m)$ to be a spectral function of function*

$$f(z) = F\left(e^{-\xi \cdot y} g(\xi)\right)(x)$$

which is analytical in domain $T(\operatorname{ch} C)$, it is necessary and sufficient that $e^{-\xi \cdot y} g(\xi) \in S'(\mathbb{R}^m)$, $\forall y \in C$.

Since we are speaking of analytical functions, certainly the question arises of the existence of their boundary values.

Proposition 2.2.3. *Let function $f(z)$ be analytical in the tube domain $T(C_R) = \mathbb{R}^m + iC_R$, where $C_R = C \cap U(0;R)$, $U(0;R)$ is the ball with center at 0 and radius R, C is a connected cone, for every number $R' < R$, and cone $C' \subset\subset C$, satisfies the estimate*

$$|f(x+iy)| \leq c(R', C')|y|^{-\alpha}(1+|x|)^{\beta},$$
$$z \in \mathbb{R}^m + i\left[C' \cap U(0;R')\right],$$

and the numbers $\alpha \geq 0$ and $\beta \geq 0$ don't depend on R' and C'.

Then there is a unique boundary value in $S'(\mathbb{R}^m)$

$$f(x) = \lim_{y \to 0,\ y \in C} f(x+iy)$$

not depending on sequence $y \to 0$, $y \in C$ (and this sequence is contained in some cone $C' \subset\subset C$).

2.3. CLASSES OF ANALYTICAL FUNCTIONS AND "EDGE OF THE WEDGE" THEOREM

Definition 2.3.1. *Function $f(z)$ belongs to class $H_p(a;C)$ $(p \geq 1, a \geq 0)$ if it is analytical in the tube cone $T(C)$ and, if for every cone $C' \subset\subset C$, it satisfies the estimate*

$$|f(z)| \leq M(C')(1+|z|)^{\beta}\left(1+|y|^{-\alpha}\right)e^{a|y|^p}, \quad z \in T(C'),$$

for some $\alpha \geq 0$, $\beta \geq 0$, not depending on C'.

Let us denote

$$H_p(a+\varepsilon, C) = \bigcap_{a'>a} H_p(a', C); \quad H_0(C) = H_1(0; C).$$

If we suppose that C consists of connected components which are the cones $C_1, C_2, \ldots$, then according to Proposition 2.2.3 the function $f(z) \in H_p(a; C)$ has boundary values

$$\lim_{y\to 0,\ y\in C_k} f(x+iy) = f_k(x) \in S'(\mathbb{R}^m), \quad k = 1, 2, \ldots,$$

and corresponding spectral functions are also from $S'(\mathbb{R}^m)$.

Functions from $H_p(a; C)$ class are well adapted to studying of relations between the growth of a function and the properties of its spectral function. We shall describe these relations in a series of following assertions.

Proposition 2.3.1. *Let function $f(z) \in H_p(a+\varepsilon; C)$, where C is a connected cone, $p > 1, a > 0$. Then its spectral function $g(\xi)$ will be represented by the sum of a finite number of derivatives from continuous functions $G_\ell(\xi)$ of power growth*

$$g(\xi) = \sum_\ell \mathcal{D}^\ell G_\ell(\xi),$$

which satisfy for all $\xi \in \overset{\prime}{\underset{*}{C}}$, $\overset{\prime}{\underset{*}{C}} \subset\subset \underset{*}{C}$ and $\varepsilon > 0$, the inequality

$$|G_\ell(\xi)| \leq M'_\varepsilon\left(\overset{\prime}{\underset{*}{C}}\right) e^{-(a'-\varepsilon)}\left[\mu_C(\xi)\right]^{p'},$$

where the numbers p, a are related to p', a' by the equations

$$\frac{1}{p} + \frac{1}{p'} = 1, \quad (p'a')^p (pa)^{p'} = 1.$$

Conversely, if $g(\xi)$ satisfies the conditions above for some numbers $a' > 0, p' > 1$ and cone $\underset{*}{C}$, then all derivatives $\mathcal{D}^\alpha f(z)$ of its Fourier-Laplace transform belong to $H_p\left(\rho_C^p a + \varepsilon,\ \operatorname{ch} C\right)$ class.

The substance of proposition 2.3.1 is very deep and very sharp. But in a such generality it is hard to use. It is convenient to apply

Corollary 2.3.1. If $f(z) \in H_p(a+\varepsilon; C)$, then $\mathcal{D}^\alpha f(z) \in H_p\left(\rho_C^p a + \varepsilon; C\right)$.

More important for our goals, however, will be the following

Proposition 2.3.2. *Let $f(z) \in H_1(a+\varepsilon; C)$ where C is a connected cone, and $a \geq 0$. Then its spectral function $g(\xi) \in S'(\mathbb{R}^m)$, and*

$$g(\xi) = 0, \quad \mu_C(\xi) > a.$$

Conversely, if $g \in S'(\mathbb{R}^m)$ vanishes in domain $\mu_C(\xi) > a$ for some $a \geq 0$ and cone C, then all derivatives of its Fourier-Laplace transform $f(z)$ belong to $H_1(\rho_C a; \operatorname{ch} C)$.

Corollary 2.3.2. *If $f(z) \in H_1(a+\varepsilon; C)$ then $\mathcal{D}^{\alpha} f(z) \in H_1(\rho_C a, \operatorname{ch} C)$. Particularly if C is a convex cone, then $H_1(a+\varepsilon, C) = H_1(a, C)$.*

An important tool for achieving our current goals is the well-known "edge of the wedge" theorem. We will not give the theorem itself here but will use one of its important and retain our reference to the theorem's title.

Definition 2.3.2. *Function $f(z)$ is called* **entire** *if it is analytical in $\mathbb{C}^m$.*

Proposition 2.3.3. *Let $f(z) \in H_p(a+\varepsilon, C)$ where the cone C consists of connected components $C_1, C_2, \ldots, C_t$, and $\operatorname{ch} C = \mathbb{R}^m$. Let further its boundary values $f_k(x) \in S'(\mathbb{R}^m)$, $k = 1, 2, \ldots, t$, coincide. Then $f(z)$ is an entire function satisfying the inequalities*

$$|f(z)| \le \begin{cases} M_\varepsilon (1+|z|)^{\beta} e^{(a+\varepsilon)\rho_C^p |y|^p}, & p > 1,\ a > 0 \\ M(1+|z|)^{\beta} e^{a\rho_C |y|}, & p = 1,\ a \ge 0\,. \end{cases}$$

Corollary 2.3.3. *If, in assumptions of Proposition 2.3.3, $p = 1$, $a = 0$, then $f(z)$ is polynomial.*

This is the multi-dimensional Liouville theorem.

3. Sobolev-Slobodetskii spaces

3.1. $H_s(\mathbb{R}^m)$ - SPACES

Let s be an arbitrary real number. By definition the Sobolev-Slobodetskii space $H_s(\mathbb{R}^m)$ consists of distributions u for which their Fourier transforms are locally integrable functions $\tilde{u}(\xi)$ such that

$$||u||_s^2 = \int_{\mathbb{R}^m} |\tilde{u}(\xi)|^2 (1+|\xi|)^{2s} d\xi < +\infty. \tag{3.1.1}$$

We will denote the Fourier image of the space $H_s(\mathbb{R}^m)$ by $\tilde{H}_s(\mathbb{R}^m)$. $\tilde{H}_s(\mathbb{R}^m)$, and, consequently $H_s(\mathbb{R}^m)$, are Hilbert spaces with respect to the inner product

$$\langle u, v\rangle_s = \int_{\mathbb{R}^m} \tilde{u}(\xi)\overline{\tilde{v}(\xi)}(1+|\xi|)^{2s} d\xi,$$

and the formula (3.1.1) defines the norm in the spaces $H_s(\mathbb{R}^m)$ and $\tilde{H}_s(\mathbb{R}^m)$.

If $s = 0$ then $\tilde{H}_0(\mathbb{R}^m) = L_2(\mathbb{R}^m)$, and by virtue of Plancherel's theorem $H_0(\mathbb{R}^m) = F^{-1}\tilde{H}_0(\mathbb{R}^m) = L_2(\mathbb{R}^m)$.

In the case $s = n$ ($n > 0$, n integer) $H_n(\mathbb{R}^m)$ consists of functions $u(x)$ that are integrable with their square functions, for which their generalized derivatives $\mathcal{D}^k u(x)$ under $1 \le |k| \le n$ are integrable with their square functions also. The norm (3.1.1) in this case is equivalent to the following norm:

$$||u||_n^2 = \sum_{|k|\le n} \int_{\mathbb{R}^m} |\mathcal{D}^k u(x)|^2 dx = \sum_{|k|\le n} \frac{1}{(2\pi)^m} \int_{\mathbb{R}^m} |\xi^k \tilde{u}(\xi)|^2 d\xi. \tag{3.1.2}$$

In the case $s = -n$, $n > 0$, n integer, the distributions from $H_{-n}(\mathbb{R}^m)$ are derivatives of functions from $L_2(\mathbb{R}^m)$ whose order is no more than n.

Proposition 3.1.1. *The functions from $C_0^\infty(\mathbb{R}^m)$ are dense in $H_s(\mathbb{R}^m)$ with respect to norm (3.1.1).*

By virtue of Proposition 3.1.1, one can define the space $H_s(\mathbb{R}^m)$ as a completion of $C_0^\infty(\mathbb{R}^m)$ (class of infinitely differentiable functions with compact support) with respect to norm (3.1.1). In the case when $s > 0$ is non-integer, one can add the formula (3.1.2), namely, the following holds.

Lemma 3.1.1. Let $0 < \lambda < 1$. The norm (3.1.1) is equivalent to the norm

$$||u||_\lambda'^2 = \int\limits_{\mathbb{R}^m}\int\limits_{\mathbb{R}^m} \frac{|u(x+y)-u(x)|^2}{|y|^{m+2\lambda}} dx dy + \int\limits_{\mathbb{R}^m} |u(x)|^2 dx.$$

If $s = n + \lambda$, $n \geq 0$, n is integer, $0 < \lambda < 1$, then the norm $||u||_{m+\lambda}$ by virtue of (3.1.2) and Lemma 3.1.1 is equivalent to the norm

$$||u||_\lambda'^2 = \sum_{|k|=n} \int\limits_{\mathbb{R}^m}\int\limits_{\mathbb{R}^m} \frac{\left|\mathcal{D}_x^k u(x+y) - \mathcal{D}_x^k u(x)\right|^2}{|y|^{m+2\lambda}} dx dy + \int\limits_{\mathbb{R}^m} |u(x)|^2 dx.$$

Let $x' = (x_1, \ldots, x_{m-1}) \in \mathbb{R}^{m-1}$. Denote by $[v]_s$ the norm in the space $H_s(\mathbb{R}^{m-1})$:

$$[v]_s = \int\limits_{\mathbb{R}^{m-1}} |\tilde{v}(\xi')|^2 (1 + |\xi'|)^{2s} d\xi'.$$

Proposition 3.1.2. *Let $s > 1/2$. Then any function $u(x', x_m) \in H_s(\mathbb{R}^m)$ is continuous function of $x_m \in \mathbb{R}$ for which its values are contained in $H_{s-1/2}(\mathbb{R}^{m-1})$.*

The following inequality holds

$$\max_{x_m \in \mathbb{R}} [u(x', x_m)]_{s-1/2} \leq c||u||_s, \quad \forall u \in H_s(\mathbb{R}^m),$$

therefore the restriction operator on the hyperplane x_m = const is a bounded operator from $H_s(\mathbb{R}^m)$ into $H_{s-1/2}(\mathbb{R}^{m-1})$.

Let us denote the restriction operator on the hyperplane $x_m = 0$ by

$$p'u(x', x_m) = u(x', 0),$$

and by Π' the following operator

$$\Pi'\tilde{u}(\xi', \xi_m) = \frac{1}{2\pi} \int\limits_{-\infty}^{+\infty} \tilde{u}(\xi', \xi_m) d\xi_m.$$

The relation between p' and Π' by Fourier transform is the following:

$$p'\tilde{w}(\xi', x_m) = \Pi'\tilde{u}(\xi, \xi_m), \tag{3.1.3}$$

where $\tilde{w}(\xi', x_m) = F_{x'\to\xi'} u(x', x_m)$, $\tilde{u}(\xi) = F_{x\to\xi} u(x)$.

Let $r \geq 0$ be integer. Denote $C_0^r(\mathbb{R}^m)$ the space of functions which are continuous together with their derivatives up to order r and such that $\lim\limits_{|x|\to\infty} u(x) = 0$, $\lim\limits_{|x|\to\infty} \mathcal{D}^p u(x) = 0$ under $|p| \leq r$.

The norm in the space $C_0^r(\mathbb{R}^m)$ is defined by formula $\sum\limits_{|p|=0}^{r} \max\limits_{x\in\mathbb{R}^m} |\mathcal{D}^p u(x)|$.

The following imbedding theorem of S.L.Sobolev holds.

Proposition 3.1.3. *Let $s > m/2$, $r < s - m/2$. Then*

$$\sum_{|p|=0}^{r} \max_{x\in\mathbb{R}^m} |\mathcal{D}^p u(x)| \leq c\|u\|_s, \quad \forall u \in C_0^\infty(\mathbb{R}^m).$$

Therefore, $H_s(\mathbb{R}^m) \subset C_0^\infty(\mathbb{R}^m)$.

Proposition 3.1.3 implies that if s is large enough then all functions from $H_s(\mathbb{R}^m)$ are smooth enough, and if $u(x) \in H_s(\mathbb{R}^m)$ for all s, then u is infinitely differentiable.

3.2. FUNCTIONS FROM $H_s(\mathbb{R}^m)$ WITH SUPPORT IN A HALF-SPACE

Let us denote $\mathbb{R}^m_\pm = \{x \in \mathbb{R}^m : \pm x_m > 0\}$.

Definition 3.2.1. *The space $H_s^\pm(\mathbb{R}^m)$ is a subspace in $H_s(\mathbb{R}^m)$ consisting of functions $u(x)$ for which their support is contained in the closed half-space $\overline{\mathbb{R}^m_\pm}$.*

Proposition 3.2.1. *The space $H_s^\pm(\mathbb{R}^m)$ is complete.*

Let us denote by $C_0^\infty(D)$ the space of infinitely differentiable functions with compact support which is contained in $\overline{D}$.

Proposition 3.2.2. *The function class $C_0^\infty(\mathbb{R}^m_\pm)$ is dense in $H_s^\pm(\mathbb{R}^m)$ with respect to norm (3.1.1).*

The key result on spaces $H_s^\pm(\mathbb{R}^m)$ is the relation between location of a function's support and analyticity of its Fourier image. One usually calls it the Wiener-Paley theorem.

Proposition 3.2.3. *Let $u \in H_s^+(\mathbb{R}^m)$. Then $\tilde{u}(\xi', \xi_m + i\tau) = F\left(ue^{-x_m\tau}\right)$ is a continuous function on τ, for $\tau \geq 0$, whose values are contained in $\tilde{H}_s(\mathbb{R}^m)$. Under almost all $\xi' \in \mathbb{R}^{m-1}$ the function $\tilde{u}(\xi', \xi_m + i\tau)$ is analytical on $\xi_m + i\tau$ in the half-space $\tau > 0$.*

The inequality holds

$$\int_{\mathbb{R}^m} |\tilde{u}(\xi', \xi_m + i\tau)|^2 (1 + |\xi'| + |\xi_m| + \tau)^{2s} d\xi' d\xi_m \leq C, \quad \tau \geq 0, \tag{3.2.1}$$

where C doesn't depend on τ.

Conversely, let τ be given locally integrable in $\mathbb{R}^m \times (0; +\infty)$ function $\tilde{u}(\xi', \xi_m + i\tau)$ satisfying the estimate (3.2.1) and analytic on $\xi_m + i\tau$ for almost all $\xi \in \mathbb{R}^{m-1}$. Then there is a function $u \in H_s^+(\mathbb{R}^m)$ such that $\tilde{u}(\xi', \xi_m + i\tau) = F\left(ue^{-x_m\tau}\right)$.

Spaces $H_s^{\pm}(\mathbb{R}^m)$ nearly adjoin other type of similar spaces. Let us describe them.

Let us denote $H_s(\mathbb{R}^m_{\pm})$ the spaces of distributions $u \in S'(\mathbb{R}^m_{\pm})$ admitting continuation ℓu on $\mathbb{R}^m$ which belongs to $H_s(\mathbb{R}^m)$. The norm in $H_s(\mathbb{R}^m_+)$ is defined as

$$||u||_s^+ = \inf_\ell ||\ell u||_s,$$

where infimum is taken on all continuations u belonging to $H_s(\mathbb{R}^m)$.

Note that the space $H_s(\mathbb{R}^m_+)$ is isomorphic to the factor-space $H_s(\mathbb{R}^m)/H_s^-(\mathbb{R}^m)$.

Proposition 3.2.4. *Let $f \in H_{-s}(\mathbb{R}^m_+)$, $u_+ \in H_s^+(\mathbb{R}^m)$, $\ell f \in H_{-s}(\mathbb{R})$ be arbitrary continuation of f. Then the form*

$$(u_+, \ell f)_0 = \int_{\mathbb{R}^m} \tilde{u}_+(\xi)\overline{\tilde{\ell f}(\xi)}d\xi \tag{3.2.2}$$

doesn't depend on the choice of continuation ℓf.

Proposition 3.2.4 and the formula (3.2.2) permit as to describe spaces conjugated to $H_s^+(\mathbb{R}^m)$.

Proposition 3.2.5. *Let $\left(H_s^+(\mathbb{R}^m)\right)^*$ be the space conjugated to $H_s^+(\mathbb{R}^m)$, s is an arbitrary number. Then $\left(H_s^+(\mathbb{R}^m)\right)^*$ is isomorphic to $H_{-s}(\mathbb{R}^m_+)$, and the value of functional $f \in H_{-s}(\mathbb{R}^m_+)$ on the element $u_+ \in H_s^+(\mathbb{R}^m)$ is given by formula (3.2.2).*

Analogously, $\left(H_{-s}(\mathbb{R}^m_+)\right)^*$ is isomorphic to $H_s^+(\mathbb{R}^m)$ under arbitrary s, and the value of functional $f \in H_{-s}(\mathbb{R}^m_+)$ on the element $f \in H_{-s}(\mathbb{R}^m_+)$ is given by formula

$$(\ell f, u_+)_0 = \int_{\mathbb{R}^m} \tilde{\ell f}(\xi)\overline{\tilde{u}_+(\xi)}d\xi = \overline{(u_+, \ell f)_0}.$$

3.3. SPACES RELATED TO CONES

Let us denote $\mathcal{L}_s^2$ the Hilbert space consisting of all functions $g(\xi)$ with finite norm

$$||g||_{(s)} = \left[\int_{\mathbb{R}^m} |g(\xi)|^2(1 + |\xi|)^{2s}d\xi\right]^{1/2} = ||g(\xi)\left(1 + |\xi|^2\right)^{s/2}||.$$

Let us denote $\mathcal{H}_s$ the set of all distributions $f(x)$ which are the Fourier transforms of functions from $\mathcal{L}_s^2$, $f = F[g]$, with the norm

$$||f||_s = ||F^{-1}[f]||_{(s)} = ||g||_{(s)}. \tag{3.3.1}$$

Let C be a convex sharp open cone, $a \geq 0$, $s \in \mathbb{R}$.

Let us denote $H_a^{(s)}(C)$ the Banach space consisting of functions $f(z)$ which are holomorphic in $T(C)$ with the norm

$$||f||_a^{(s)} = \sup_{y \in C} e^{-a|y|}||f(x+iy)||_s \tag{3.3.2}$$

and denote $H_0^{(s)}(C) = H^{(s)}(C)$, $||\cdot||_0^{(s)} = ||\cdot||^{(s)}$.

Proposition 3.3.1. *The function $f(z)$ belongs to $H_a^{(s)}(C)$ if and only if its spectral function $g(\xi)$ belongs to the class $\mathcal{L}_s^2\left(\tilde{C}_a\right)$, where $\tilde{C}_a = \{\xi : \mu_c(\xi) \leq a\}$.*

The following equalities are valid under these assumptions:

$$||f||_a^{(s)} = ||f_+|| = ||g||_{(s)} \tag{3.3.3}$$

where $f_+(x)$ is a boundary value in $\mathcal{H}_s$ of function $f(z)$, $z = x + iy$, under $y \to 0$, $y \in C$, and $f_+ = F[g]$.

Corollary 3.3.1. *The spaces $H_a^{(s)}(C)$ and $\mathcal{L}_s^2\left(\tilde{C}_a\right)$ are (linear) isomorphic and isometric, and this isomorphism is realized by a Fourier-Laplace transform.*

Corollary 3.3.2. *Any function $f(z) \in H_a^{(s)}(C)$ has, for $y \to 0$, $y \in C$ a (unique) boundary value $f_+(x) \in \mathcal{H}_s$, and the correspondence $f \to f_+$ is an isometry.*

Corollary 3.3.3. *If the cone C is non-sharp, and $f \in H_0^{(s)}(C)$* then $f(z) \equiv 0$.

4. Pseudodifferential operators and equations in a half-space

4.1. PSEUDODIFFERENTIAL OPERATORS

Let $A(\xi)$ be a locally integrable function satisfying the inequality

$$|A(\xi)| \leq C(1+|\xi|)^{\alpha}. \tag{4.1.1}$$

We will denote the class of such functions by

Definition 4.1.1. *By pseudodifferential operator we mean the operator which is defined on functions from $S(\mathbb{R}^m)$ by the formula*

$$(Au)(x) = \frac{1}{2\pi}\int_{\mathbb{R}^m} A(\xi)\tilde{u}(\xi)e^{-ix\cdot\xi}d\xi.$$

The function $A(\xi)$ is called a symbol of operator A.

If $A(\xi)$ is polynomial on ξ, i.e., $a(\xi) = \sum\limits_{|k|\leq n} a_k\xi^k$, then by virtue of a Fourier transform

$$(Au)(x) = \frac{1}{(2\pi)^m}\int_{\mathbb{R}^m}\sum_{|k|\leq n} a_k\xi^k\tilde{u}(\xi)e^{-ix\cdot\xi}d\xi = \sum_{|k|\leq n} a_k\mathcal{D}^k u(x),$$

i.e., A is a differential operator. Therefore the class of pseudodifferential operators includes the class of differential operators.

Example 4.1.1. We denote Y_k, $1 \leq k \leq m$, a pseudodifferential operator with symbol $\eta_k = \frac{\xi_k}{|\xi|}$ (this symbol represents a homogeneous function of order 0, i.e., it is symbol of some multidimensional singular integral operator of Calderon-Zygmund [49,171,172]). Then

$$(Y_k u)(x) = -i\frac{\Gamma\left(\frac{m+1}{2}\right)}{\pi^{\frac{m+1}{2}}}\ v.p. \int\limits_{\mathbb{R}^m} \frac{x_k - y_k}{|x-y|^{m+1}} u(y)dy,$$

where Γ is the Euler Γ - function, and v.p. denotes the following:

$$v.p. \int\limits_{\mathbb{R}^m} \frac{x_k - y_k}{|x-y|^{m+1}} u(y)dy = \lim_{\varepsilon \to 0} \int\limits_{\mathbb{R}^m \setminus B_\varepsilon(x)} \frac{x_k - y_k}{|x-y|^{m+1}} u(y)dy,$$

$B_\varepsilon(x)$ is the ball with center x and radius ε.

Multidimensional singular integral operators Y_k a called Riesz operators.

Example 4.1.2. Let Λ be pseudodifferential operator with symbol $|\xi|$. Applying the expansion

$$|\xi| = \sum_{k=1}^{m} \xi_k \frac{\xi}{|\xi|},$$

the properties of the Fourier transform and example 4.1.1., Λ can be represented in the form

$$(\Lambda u)(x) = \sum_{k=1}^{m} Y_k \mathcal{D}^k u =$$

$$= \sum_{k=1}^{m} \frac{\Gamma\left(\frac{m+1}{2}\right)}{\pi^{\frac{m+1}{2}}} \int\limits_{\mathbb{R}^m} \frac{x_k - y_k}{|x-y|^{m+1}} \frac{\partial u(y)}{\partial y_k} dy.$$

The operator Λ is called an integro-differential Calderon-Zygmund operator.

Let us describe some additional important properties of pseudodifferential operators.

Proposition 4.1.1. *Let $A(\xi) \in S^0_\alpha$. Then a pseudodifferential operator A with symbol $A(\xi)$ satisfies, under arbitrary s, the estimate*

$$||Au||_{s-\alpha} \leq C||u||_s, \quad \forall u \in S(\mathbb{R}^m), \tag{4.1.2}$$

and therefore extends its continuity up to a bounded operator acting from $H_s(\mathbb{R}^m)$ into $H_{s-\alpha}(\mathbb{R}^m)$.

Note that if $A(\xi)$ is an unbounded function in a neighbourhood of some point, estimate (4.1.2) does not hold.

Denote via $\tilde{H}^{\pm}_s(\mathbb{R}^m)$ the Fourier-image of the space $H^{\pm}_s(\mathbb{R}^m)$.

Proposition 4.1.2. *Let the function $A_+(\xi', \xi_m + i\tau)(A_-(\xi', \xi_m + i\tau))$ be continuous on a set of variables under $\xi' \neq 0$, $\tau \geq 0$, $(\tau \leq 0)$, analytical on $\xi_m + i\tau$ under $\tau > 0$ $(\tau < 0)$ and satisfy the inequality*

$$|A_\pm(\xi', \xi_m + i\tau)| \leq c(1 + |\xi'| + |\xi_m| + |\tau|)^\alpha. \tag{4.1.3}$$

Then the multiplication operator on $A_\pm(\xi',\xi_m)$ is bounded as an operator acting from $\tilde{H}_s^\pm(\mathbb{R}^m)$ into $H_{s-\alpha}^\pm(\mathbb{R}^m)$.

Proposition 4.1.2 plays a very important role in the study of pseudidifferential equations in half-space (see below).

4.2. CAUCHY TYPE INTEGRALS AND SYMBOL FACTORIZATION

For a function $\tilde{f}(\xi',\xi_m) \in S(\mathbb{R}^m)$ we introduce

$$F_+(\xi',\xi_m+i\tau) = \frac{i}{2\pi}\int\limits_{-\infty}^{+\infty} \frac{\tilde{f}(\xi',\eta_m)}{\xi_m+i\tau-\eta_m} d\eta_m, \quad \tau > 0. \tag{4.2.1}$$

Proposition 4.2.1. For arbitrary $\tilde{f}(\xi',\xi_m) \in S(\mathbb{R}^m)$, the function $F_+(\xi',\xi_m+i\tau)$ is analytical on $\xi_m+i\tau$ in half-space $\tau > 0$, and there exists is $F_+(\xi',\xi_m+i0) = \lim\limits_{\tau\to+0} F_+(\xi',\xi_m+i\tau)$ and

$$F_+(\xi',\xi_m+i\tau) = \frac{1}{2}\tilde{f}(\xi',\xi_m) + \text{ v.p. } \frac{i}{2\pi}\int\limits_{-\infty}^{+\infty} \frac{\tilde{f}(\xi',\eta_m)}{\xi_m-\eta_m} d\eta_m, \tag{4.2.2}$$

$$\text{where v.p.} \int\limits_{-\infty}^{+\infty} \frac{\tilde{f}(\xi',\eta_m)}{\xi_m-\eta_m} d\eta_m = \lim_{\varepsilon\to 0} \int\limits_{|\xi_m-\eta_m|>\varepsilon} \frac{\tilde{f}(\xi',\eta_m)}{\xi_m-\eta_m} d\eta_m$$

is the integral in the sense of a Cauchy pricipal value.

Analogously, if

$$F_-(\xi',\xi_m+i\tau) = -\frac{i}{2\pi}\int\limits_{-\infty}^{+\infty} \frac{\tilde{f}(\xi',\eta_m}{\xi_m+i\tau-\eta_m} d\eta_m, \quad \tau < 0, \tag{4.2.3}$$

$\tilde{f}(\xi',\xi_m) \in S(\mathbb{R}^m)$, then $F_-(\xi',\xi_m+i\tau)$ is analytical on $\xi_m+i\tau$ in the half-space $\tau < 0$, the exists $F_+(\xi',\xi_m-i0) = \lim\limits_{\tau\to-0} F_-(\xi',\xi_m+i\tau)$, and

$$F_-(\xi',\xi_m-i0) = \frac{1}{2}\tilde{f}(\xi',\xi_m) - \text{ v.p. } \frac{i}{2\pi}\int\limits_{-\infty}^{+\infty} \frac{\tilde{f}(\xi',\eta_m)}{\xi_m-\eta_m} d\eta_m. \tag{4.2.4}$$

Let us denote

$$\Pi^\circ \tilde{f} = \text{ v.p. } \frac{i}{2\pi}\int\limits_{-\infty}^{+\infty} \frac{\tilde{f}(\xi',\eta_m)}{\xi_m-\eta_m} d\eta_m,$$

$$\Pi^+ \tilde{f} = F_+(\xi', \xi_m + i0) = \frac{i}{2\pi} \int_{-\infty}^{+\infty} \frac{\tilde{f}(\xi', \eta_m)}{\xi_m + i0 - \eta_m} d\eta_m,$$

$$\Pi^- \tilde{f} = F_-(\xi', \xi_m - i0) = -\frac{i}{2\pi} \int_{-\infty}^{+\infty} \frac{\tilde{f}(\xi', \eta_m)}{\xi_m - i0 - \eta_m} d\eta_m.$$

According to (4.2.2) and (4.2.4)

$$\Pi^+ \tilde{f} = \frac{1}{2}\tilde{f} + \Pi^0 \tilde{f}, \quad \Pi^- \tilde{f} = \frac{1}{2}\tilde{f} - \Pi^0 \tilde{f},$$

and therefore

$$\tilde{f} = \Pi^+ \tilde{f} + \Pi^- \tilde{f}$$

for any function $\tilde{f} \in S(\mathbb{R}^m)$.

The role of a Cauchy type integral for forthcoming study of equations in half-space is shown in the following proposition.

Let Θ^+ be a multiplication operator on a function $\Theta(x_m)$ which is equal to 1 under $x_m > 0$ and 0 under $x_m < 0$.

Proposition 4.2.2. *For arbitrary $f \in S(\mathbb{R}^m)$*

$$F\left(\Theta^+ f(x', x_m)\right) = \Pi^+ \tilde{f}, \tag{4.2.5}$$

where $\tilde{f} = Ff$ is the Fourier transform of function $f(x', x_m)$.

Analogously, if one introduces Θ^- as a multiplication operator on $(1 - \Theta(x_m))$, then

$$F(\Theta^- f) = \Pi^- \tilde{f}, \quad \forall f \in S(\mathbb{R}^m). \tag{4.2.6}$$

Proposition 4.2.3. *The operator $\Pi^\pm$ defined on $S(\mathbb{R}^m)$ by formulas (4.2.1) and (4.2.3) are bounded with respect to norm $\tilde{H}_\delta(\mathbb{R}^m)$ under $|\delta| < 1/2$, and therefore they can be extended in continuity on $\tilde{H}_\delta(\mathbb{R}^m)$ as bounded operators acting from $\tilde{H}_\delta(\mathbb{R}^m)$ into $\tilde{H}^\pm_\delta(\mathbb{R}^m)$.*

However, note that for $\delta = \pm 1/2$ these operators are unbounded in $\tilde{H}_\delta(\mathbb{R}^m)$.

Now we consider symbol factorization.

Let symbol $A(\xi)$ is infinite differentiable for $\xi \neq 0$ and $A(\xi)$ be homogeneous function of order $\alpha + i\beta$, i.e.,

$$A(t\xi) = t^{\alpha + i\beta} A(\xi), \quad \forall t = 0.$$

We will denote the class of such symbols by $C^\infty_{\alpha+i\beta}$.

Definition 4.2.1. *Symbol $A(\xi)$ is called elliptic if*

$$A(\xi) \neq 0, \quad \forall \xi \neq 0. \tag{4.2.7}$$

Definition 4.2.2. *By a homogeneous factorization of the elliptic symbol $A(\xi)$ in the variable ξ_m we mean a resresentation of $A(\xi', \xi_m)$ in the form*

$$A(\xi', \xi_m) = A_+(\xi', \xi_m) A_-(\xi', \xi_m), \tag{4.2.8}$$

where $A_+(\xi', \xi_m)$ and $A_-(\xi', \xi_m)$ satisfy the following conditions:

a) $A_+(\xi', \xi_m)$ under $\xi' \neq 0$ admits an analytical continuation into the upper half-plane $\tau > 0$;

b) $A_+(\xi', \xi_m)$ is continuous on a set of variables (ξ', ξ_m, τ) under $\tau \geq 0$, $|\xi'| + |\xi_m| + \tau > 0$;

c) $A_+(\xi', \xi_m + i\tau)$ is a homogeneous function of order $æ = æ_1 + iæ_2$:

$$A_+(t\xi', t(\xi_m + i\tau)) = t^{æ} A_+(\xi', \xi_m + i\tau), \quad \forall t > 0;$$

d) $A_+(\xi', \xi_m + i\tau) \neq 0$ for $\tau \geq 0$, $|\xi'| + |\xi_m| + \tau > 0$.

Analogously, the function $A_-(\xi', \xi_m)$ admits for $\xi' \neq 0$ an analytical continuation on ξ_m into the lower half-space $\tau < 0$, it is continuous and does not vanish under $\tau \leq 0$, $|\xi'| + |\xi_m| + \tau > 0$, and its order of homogeneity of $A_-(\xi', \xi_m + i\tau)$ on $(\xi', \xi_m + i\tau)$ is equal to $\alpha + i\beta - æ$.

Since multiplying A_+ by a constant and dividing A_- by the same constant we will obtain another factorization having all the above properties, then usually we add the normalization condition (to rule out this situation):

$$A_+(0; 1) = 1. \tag{4.2.9}$$

Proposition 4.2.4. *Let $A(\xi)$ be elliptic symbol. Then $A(\xi)$ admits a unique homogeneous factorization under normalization condition (4.2.9).*

Note that such a factorization, with the help of a Cauchy type integral, results in an explicit constant

Definition 4.2.3. *The order of homogeneity $æ = æ_1 + æ_2$ of a function $A_+(\xi', \xi_m)$ is called a factorization index of the elliptic symbol $A(\xi', \xi_m)$.*

As we will see below, the value of a factorization index plays an exceptionally important role in the solvability theory of pseududifferential equations in half-space.

4.3. EQUATIONS IN A HALF-SPACE

Let A be a pseudodifferential operator with symbol $A(\xi) \in C^{\infty}_{\alpha+i\beta}$. Let

$$\hat{A}(\xi', \xi_m) = A((1 + |\xi'|)\omega, \xi_m), \tag{4.3.1}$$

where $\omega = \frac{\xi'}{|\xi'|}$. By changing symbol A to symbol $\hat{A}$, we stipulate that symbol $\hat{A}$ has better properties; in particular, it is bounded for $\alpha < 0$, and the difference $A - \hat{A}$, roughly speaking, does not affect the character of solvability of the equation. Therefore by pseudodifferential equation in half-space $\mathbb{R}^m_+$ we mean a equation of type

$$p\hat{A}u_+ = f, \tag{4.3.2}$$

where solution u_+ is sought in the space $H^+_s(\mathbb{R}^m)$, $f \in H_s(\mathbb{R}^m_+)$, p is the restriction operator on $\mathbb{R}^m_+$.

Everywhere below in this section we assume that $A(\xi) \in C^{\infty}_{\alpha+i\beta}$ is an elliptic symbol, that

$$A(\xi', \xi_m) = A_+(\xi', \xi_m)A_-(\xi', \xi_m)$$

is a homogeneous factorization of $A(\xi', \xi_m)$, and that æ is a factorization index of $A(\xi', \xi_m)$.

Let us assume that $\mathrm{Re}\,æ - s \neq 1/2 \pmod k$, k an arbitrary integer. Then obviously there is such integer n an that

$$\mathrm{Re}\,æ - s = n + \delta, \quad |\delta| < 1/2. \tag{4.3.3}$$

It is proved that in dependence on $\mathrm{sign}\, n$ the solvability pictures for equation (4.3.2) strongly differ from one another.

Proposition 4.3.1. *Let $n = 0$. Then the equation (4.3.2) for any right-hand side $f \in H_{s-\alpha}(\mathbb{R}^m_+)$ has a unique solution $u_+ \in H^+_s(\mathbb{R}^m)$ for which its Fourier transform is given by the formula*

$$\tilde{u}_+(\xi) = \hat{A}_+^{-1}\Pi^+\hat{A}_-^{-1}\widetilde{\ell f}, \tag{4.3.4}$$

where ℓf is the arbitrary continuation of f on $\mathbb{R}^m$. The a priori estimate holds

$$||u_+||_s \leq c||f||^+_{s-\alpha}.$$

The result (4.3.4) does not depend on the choice of continuation ℓf.

Proposition 4.3.2.*Let $n > 0$. Then the equation (4.3.2) for any right-hand side $f \in H_{s-\alpha}(\mathbb{R}^m_+)$ has a solution $u_+(x) = F^{-1}\tilde{u}_+(\xi)$, where*

$$\tilde{u}_+(\xi) = \hat{A}_+^{-1}\hat{Q}\Pi^+\hat{Q}^{-1}\hat{A}_-^{-1}\widetilde{\ell f} + \frac{\sum\limits_{k=1}^{n} \tilde{c}_k(\xi')\xi_m^{k-1}}{\hat{A}_+(\xi)},$$

Q is a arbitrary polynomial on ξ_m of order n. This solution is non-unique and depends on n arbitrary functions $c_k(x') \in H_{s_k}(\mathbb{R}^{m-1})$, $s_k = s - \mathrm{Re}\,æ + k - 1/2$, $1 \leq k \leq n$. The a priori estimate holds

$$||u_+||_s \leq c\left(||f||^+_{s-\alpha} + \sum_{k=1}^{n}[c_k]_{s_k}\right), \quad s_k = s - \mathrm{Re}\,æ + k - 1/2,\ 1 \leq k \leq n.$$

The last case $n < 0$ is more interesting and requires new concepts.

Definition 4.3.1. *Let $C(\xi) \in S^0_\alpha$. By a pseudodifferential operator of potential type we mean the operator which is defined by the formula*

$$C\left(v(x') \otimes \delta(x_m)\right) = F^{-1}\left(C(\xi', \xi_m)\tilde{v}(\xi')\right), \quad \forall v(x') \in C_0^\infty(\mathbb{R}^{m-1}).$$

Example 4.3.1. Let P_m be a pseudodifferential operator with symbol $\frac{\xi_m}{|\xi|^2}$, $m \geq 2$. The form of operator P_m in x - space is the following:

$$P_m u = \frac{-i\Gamma\left(\frac{m+1}{2}\right)}{2\pi^{\frac{m+1}{2}}} \int\limits_{\mathbb{R}^m} \frac{x_m - y_m}{|x-y|^m}\, dy,$$

and then we have

$$P_m\left(v(x')\otimes\sigma(x_m)\right)=-\frac{-i\Gamma\left(\frac{m+1}{2}\right)}{2\pi^{\frac{m+1}{2}}}\int\limits_{\mathbb{R}^m}\frac{x_m v(\xi')dy'}{[x_m^2+(x'-y')^2]^{1/2}},$$

i.e., $iP_m\left(v(x')\otimes\delta(x_m)\right)$ is a double layer potential for the Laplace equation.

Let us denote $\hat{\Lambda}_+(\xi',\xi_m)=\xi_m+i|\xi'|+i$.

Proposition 4.3.3. *Let $n<0$. Then for any $f\in H_{s-\alpha}(\mathbb{R}^m_+)$ there is a unique collection of functions $u_+\in H_s^+(\mathbb{R}^m)$, $v_k(x')\in H_{m_k}(\mathbb{R}^{m-1})$, $m_k=|n|-\delta-k+1/2$, $1\le k\le |n|$, such that*

$$p\hat{A}\left[u_+ +\sum_{k=1}^{|n|}\hat{A}_+^{-1}\hat{\Lambda}_+^{-k}\left(v_k(x')\otimes\delta(x_m)\right)\right]=f,$$

where $\hat{A}_+^{-1}\hat{\Lambda}_+^{-k}\left(v_k(x')\otimes\delta(x_m)\right) = F_\xi^{-1}\left[\hat{A}_+^{-1}(\xi',\xi_m)\hat{\Lambda}_+^{-k}(\xi',\xi_m)\tilde{v}_k(\xi')\right]$ are potential type operators.

Particularly, the equation (4.3.2) has a unique solution belonging to $H_s^+(\mathbb{R}^m)$ if and only if $f(x)$ for almost all ξ', satisfies the conditions

$$\int\limits_{-\infty}^{+\infty}\xi_m^{k-1}\frac{\tilde{\ell f}(xi',\xi_m)}{\hat{A}_-(\xi',\xi_m)}d\xi_m=0,\quad 1\le k\le|n|. \tag{4.3.5}$$

The following estimates hold:

$$||u_+||_s\le c||f||^+_{s-\alpha},\quad \sum_{k=1}^{|n|}[v_k]_{m_k}\le c||f||^+_{s-\alpha}.$$

As a commentary to Proposition 4.3.3 we note the conditions (4.3.5) in x - space. There they look more transparent:

$$\mathcal{D}_n^{k-1}\hat{A}_-^{-1}\ell f\Big|_{x_m=0}=0,\quad f\in H_{s-\alpha}(\mathbb{R}^m_+), \tag{4.3.6}$$

where $\hat{A}_-^{-1}$ is a pseudodifferential operator with symbol $\hat{A}_-^{-1}(\xi',\xi_m)$.

Propositions 4.3.1, 4.3.2, 4.3.3 cover all solvability cases of pseudodifferential equation (4.3.2) in half-space, also if in Proposition 4.3.1 the solution exists and is unique, then in the other two propositions we must refine the statements to determine when the solution exists and is unique. Roughly speaking, in Proposition 4.3.2 the problem is underdetermined, and we need conditions permitting us to find unknown functions $c_k(x')$, $k=1,\ldots,n$. In Proposition 4.3.3 the problem is overdetermined, and we must suggest generalized statement of the problem from which this overdetermination is missing.

4.4. BOUNDARY VALUE PROBLEMS

We begin with an analysis of Proposition 4.3.2. To define unknown functions $c_k(x')$ we add to equation (4.3.2) n boundary conditions and consider the following boundary value problem:

$$\begin{cases} p\hat{A}u_+ = f \\ p'\hat{B}_j u_+ = g_j(x'), \quad 1 \le l \le m, \end{cases} \tag{4.4.1}$$

where (we recall) $\hat{A}$ is an elliptic pseudodifferential operator with symbol $\hat{A}(\xi', \xi_m)$, $A(\xi', \xi_m) \in C^\infty_{\alpha+i\beta}$, $f \in H_{s-\alpha}(\mathbb{R}^m_+)$, B_j are pseudodifferential operators with symbols $\hat{B}_j(\xi', \xi_m)$, $B_j(\xi', \xi_m) \in C^\infty_{\beta_j}$, p' is a restriction operator on the hyperplane $x_m = 0$.

In addition (only in order to avoid complicating the exposition with technical details) we assume the following condition holds:

$$|B_j(\xi', \xi_m)| \le |\xi'|^{\operatorname{Re}\beta_j - \beta_{j1}}(|\xi'| + |\xi_m|)^{\beta_{j1}}, \quad \beta_{j1} < s - 1/2. \tag{4.4.2}$$

Let us denote

$$\hat{b}_{jk}(\xi') = \frac{1}{2\pi} \int\limits_{-\infty}^{+\infty} \frac{\hat{B}_j(\xi', \xi_m)}{\hat{A}_+(\xi', \xi_m)} d\xi_m, \quad 1 \le j, k \le n, \tag{4.4.3}$$

and the matrix with elements $\hat{b}_{jk}(\xi)$ we will denote by $||\hat{b}_{jk}(\xi')||^n_{j,k=1}$.

We have the following proposition related to unique solvability of boundary value problem (4.4.1).

Proposition 4.4.1. *Let a $A(\xi) \in C^\infty_{\alpha+i\beta}$ be elliptic symbol, æ be the factorization index of $A(\xi', \xi_m)$, $\operatorname{Re}æ - s = n + \delta$, $n > 0$ integer, $|\delta| < 1/2$. Let $B_j(\xi) \in C^\infty_{\beta_j}$, the condition (4.2.2) holds, and*

$$\det ||\hat{b}_{jk}(\xi')||^n_{j,k=1} \ne 0, \quad \forall \xi' \ne 0. \tag{4.4.4}$$

Then for arbitrary $f(x) \in H_{s-\alpha}(\mathbb{R}^m_+)$ and $g_j(x') \in H_{s-\operatorname{Re}\beta_j-1/2}(\mathbb{R}^{m-1})$, $1 \le j \le n$, there is a unique solution $u_+ \in H^+_s(\mathbb{R}^m)$ of boundary value problem (4.4.1). The a priori estimate holds

$$||u_+||_s \le c\left(||f||^+_{s-\alpha} + \sum_{j=1}^n [g_j]_{s-\operatorname{Re}\beta_j-1/2}\right).$$

The condition (4.4.4) to make an analogy to case of elliptic problems for differential equations is called the Shapiro-Lopatinskii condition.

Let us turn to analysis of Proposition 4.3.3. Since the problem in this case is "overdetermined", one has a possibility to extend the number of unknowns.

Proposition 4.3.3 requires that these unknowns have be related to potential type operators. Let $C(\xi',\xi_m) \in C^\infty_{\alpha+i\beta}$ and

$$|C(\xi',\xi_m)| \le c|\xi'|^{\alpha-\alpha'}(|\xi'|+|\xi_m|)^{\alpha_1}. \qquad (4.4.5)$$

Proposition 4.4.2. *If $s < \alpha - \alpha_1 - 1/2$ and inequality (4.4.5) holds, then potential type operator $\hat{C}(v(x') \otimes \delta(x_m))$ extends on continuity up to a bounded operator acting from $H_{s+1/2}(\mathbb{R}^{m-1})$ into $H_{s-\alpha}(\mathbb{R}^m)$.*

Indeed if $v(x') \in C_0^\infty(\mathbb{R}^{m-1})$, then by virtue of (4.4.5) we have

$$||\hat{C}(\xi',\xi_m)\tilde{v}(\xi')||^2_{s-\alpha} \le c \int\limits_{\mathbb{R}^m} (1+|\xi'|)^{2(\alpha-\alpha_1)} \times$$

$$\times (1+|\xi'|+|\xi_m|)^{2(s+\alpha_1-\alpha)}|\tilde{v}(\xi')|^2 d\xi' d\xi_m,$$

and the integral converges, because $s + \alpha_1 - \alpha < -1/2$.

Further, taking change of variables $\xi_m = (1+|\xi'|)\eta_m$ we obtain

$$||\hat{C}(\xi',\xi_m)\tilde{v}(\xi_m)||^2_{s-\alpha} \le c \int\limits_{\mathbb{R}^m} |\tilde{v}(\xi')|^2(1+|\xi'|)^{2(s+1/2)}d\xi' = c[v]^2_{s+1/2}, \qquad (4.4.6)$$

and, since the functions from class $C_0^\infty(\mathbb{R}^{m-1})$ are dense in $H_{s+1/2}(\mathbb{R}^{m-1})$ the estimate (4.4.6) is valid for all $v \in H_{s+1/2}(\mathbb{R}^{m-1})$.

The operator $p\hat{C}(v(x') \otimes \delta(x_m))$ is bounded an operator acting from $H_{s+1/2}(\mathbb{R}^{m-1})$ into $H_{s-\alpha}(\mathbb{R}^{m-1}_+)$ because the restriction operator p is bounded from $H_{s-\alpha}(\mathbb{R}^{m-1})$ into $H_{s-\alpha}(\mathbb{R}^{m-1}_+)$ for $s < \alpha - \alpha_1 - 1/2$. Otherwise, for $s \ge \alpha - \alpha_1 - 1/2$, the $p\hat{C}(v(x') \otimes \delta(x_m)) \notin H_{s-\alpha}(\mathbb{R}^m)$, generally speaking, and the problem is to introduce the class of symbols for which

$$||p\hat{C}(v(x') \otimes \delta(x_m))||^+_{s-\alpha} \le c[v]_{s+1/2}, \quad \forall v(x') \in H_{s+1/2}(\mathbb{R}^{m-1}).$$

Let symbol $C(\xi',\xi_m) \in C^\infty_{\alpha+i\beta}$ have the following property: for any $N > 0$ one has the representation:

$$C(\xi',\xi_m) = C_N^-(\xi',\xi_m) + R_N(\xi',\xi_m), \qquad (4.4.7)$$

where $C_N^-(\xi',\xi_m)$ admits analytical continuation on $\xi_m + i\tau$ into half-space $\tau < 0$, the function $C_N^-(\xi',\xi_m+i\tau)$ is continuous for $\xi' \ne 0$ and satisfies the inequality

$$|C_N^-(\xi',\xi_m+i\tau)| \le c_N(|\xi'|+|\xi_m|+|\tau|)^{\alpha_1}|\xi'|^{\alpha-\alpha_1},$$

and the function $R_N(\xi',\xi_m)$ satisfies the inequality

$$|R_N(\xi',\xi_m)| \le c_N \frac{|\xi'|^N}{(|\xi'|+|\xi_m|)^{N-\alpha}}.$$

In this case, the following proves to hold.

Proposition 4.4.3. *Let $C(\xi', \xi_m) \in C^{\infty}_{\alpha+i\beta}$ satisfying the condition (4.4.7). Then for arbitrary s the operator $p\hat{C}(v(x') \otimes \delta(x_m))$ is bounded from $H_{s+1/2}(\mathbb{R}^{m-1})$ into $H_{s-\alpha}(\mathbb{R}^m_+)$.*

Let us consider the following class of pseudodifferential equations. Let

$$p\left(\hat{A}u_+ + \sum_{k=1}^{|n|} \hat{C}_k(v_k \otimes \delta(x_m))\right) = f(x), \tag{4.4.8}$$

where $f \in H_{s-\alpha}(\mathbb{R}^m_+)$, $\hat{C}_k$ are pseudodifferential operators with symbols $\hat{C}_k(\xi', \xi_m) \in C^{\infty}_{\alpha_k+i\beta_k}$; if $s \geq \alpha - \alpha_k - 1/2$, then one assumes that $C_k(\xi', \xi_m)$ admits the expansion of type (4.4.7)

$$C_k(\xi', \xi_m) = C_k^-(\xi', \xi_m) + R_k(\xi', \xi_m), \tag{4.4.9}$$

where $C_k^-(\xi', \xi_m)$ is a minus-function satisfying the above conditions, and $R_k(\xi', \xi_m)$ satisfies the inequality

$$|R_k(\xi', \xi_m)| \leq c|\xi'|^{\alpha_k - \alpha_{k1}}(|\xi'| + |\xi_m|)^{\alpha_{k1}}, \quad \alpha_{k1} < \alpha - s - 1/2. \tag{4.4.10}$$

The functions $u_+ \in H_s^+(\mathbb{R}^m)$ and $v_k(x') \in H_{s_k}(\mathbb{R}^{m-1})$, $s_k = s = \alpha + \alpha_k + 1/2$, $1 \leq k \leq |n|$, are sought.

Let us denote

$$\hat{c}_{kj}(\xi') = \int_{-\infty}^{+\infty} \frac{\hat{R}_j(\xi', \xi_m)\xi_m^{k-1}}{\hat{A}_-(\xi', \xi_m)} d\xi_m, \quad 1 \leq k, j \leq |n|,$$

and the matrix with elements $\hat{c}_{kj}$ we denote $||\hat{c}_{kj}(\xi')||_{k_{1j}=1}^{|n|}$.

Proposition 4.4.4. Let $A(\xi)$ be an elliptic symbol, æ be factorization index of $A(\xi)$, $\mathrm{Re}\,æ - s = n + \delta$, $n < 0$ integer, $|\delta| < 1/2$.

Let $C_k(\xi) \in C^{\infty}_{\alpha_k+i\beta_k}$ and satisfy the condition (4.4.9). Then given valodity of the analogue of the Shapiro-Lopatinskii condition

$$\det ||\hat{c}_{kj}(\xi')||_{k,j=1}^{|n|} \neq 0, \quad \forall \xi' \neq 0,$$

for arbitrary $f \in H_{s-\alpha}(\mathbb{R}^m_+)$ there exists, a unique solution $u_+ \in H_s^+(\mathbb{R}^m)$, $v_k(x') \in H_{s_k}(\mathbb{R}^{m-1})$, $s_k = s - \alpha + \alpha_k + 1/2$, $1 \leq k \leq |n|$, of equation (4.4.8). The following estimates hold:

$$\sum_{j=1}^{|n|} [v_j]_{s-\alpha+\alpha_j+1/2} \leq c||f||_{s-\alpha}^+,$$

$$||u_+|| \leq c\left(||f||_{s-\alpha}^+ \sum_{j=1}^{|n|} [v_j]_{s-\alpha+\alpha_j+1/2}\right) \leq C||f||_{s-\alpha}^+.$$

With this section we complete the discussion and survey of preliminary materials. We recomended that the reader review the highlights that will be necessary to continue reading this book and to understand the constructions that will follow.

5. Wave factorization

It was clearly shown in the previous section that, in the theory of boundary value problems for elliptic pseudodifferential equations which was constructed by M.I.Vishik and G.I.Eskin, the key role was played the symbol factorization of elliptic pseudodifferential operators on one of variables with the other variables fixed. Depending on factorization index one needs to consider the different boundary value problems reflecting the singularities of appeared situation. When studying the pseudodifferential equation in cone (angle) it proves that analogous role will play the wave factorization related to exit into compex domain with respect to all variables at once. Such factorization permits one to use a multivariable variant of the Wiener-Hopf method and to obtain for a pseudodifferential operator in angle the same solvability picture which we had in the half-space case. This approach has been applied to some classes of differential operators too, but as of now it has not been useful. In this section we will introduce the concept of wave factorization for symbols of pseudodifferential operators and we will verify that the set of symbols which admit the wave factorization is large enough.

Let $A(\xi)$, $\xi \in \mathbb{R}^m$, be symbol of pseudodifferential operator A satisfying the condition

$$c_1 \leq |A(\xi)(1+|\xi|)^{-\alpha}| \leq c_2, \tag{5.1}$$

where c_1, c_2 are positive constants. The number $\alpha \in \mathbb{R}$ we will call the order of pseudodifferential operator A. It is obvious that the symbol $A(\xi)$ is elliptic, i.e., $A(\xi) \neq 0$, $\forall \xi \in \mathbb{R}^m$. The class of such symbols will be denoted C_α.

Definition 5.1. *By wave factorization of symbol $A(\xi)$ with respect to cone C_+^a* [1]) *one means its representation in the form*

$$A(\xi) = A_{\neq}(\xi)A_{=}(\xi),$$

where the factors $A_{\neq}(\xi)$, $A_{=}(\xi)$ must have the following properties:

1) *$A_{\neq}(\xi)$, $A_{=}(\xi)$ admit an analytical continuation into $T(\overset{*}{C}{}_+^a)$, $T(\overset{*}{C}{}_-^a)$ respectively, which satisfies the inequalities.*

$$|A_{\neq}^{\pm 1}(\xi + i\tau)| \leq c_1(1+|\xi|+|\tau|)^{\pm æ},$$

$$|A_{=}^{\pm 1}(\xi - i\tau)| \leq c_2(1+|\xi|+|\tau|)^{\pm(\alpha - æ)}, \quad \forall \tau \in \overset{*}{C}{}_+^a,$$

where $\overset{*}{C}{}_-^a$ is the cone opposite to cone $\overset{*}{C}{}_+^a$:

$$\overset{*}{C}{}_-^a = -\overset{*}{C}{}_+^a.$$

The number æ we will call the index of wave factorization.

The following example has a very important meaning, and we recommend that reader to pay serious attention to it.

---

[1]) see Section 2.1

Example 5.1. Let

$$A = -\frac{\partial^2}{\partial x_1^2} - \cdots - \frac{\partial^2}{\partial x_1^m} - k^2, \quad k \in \mathbb{R} \setminus \{0\},$$

and then according to some properties of the Fourier transform the symbol of this operator has the form

$$A(\xi) = \xi_1^2 + \xi_2^2 + \cdots + \xi_m^2 + k^2.$$

Let us consider the function

$$\sqrt{a^2+1}\, z_m + \sqrt{a^2 z_m^2 - z_1^2 - \cdots - z_{m-1}^2 - k^2}$$

as a function of m complex variables $z_j = \xi_j + i\eta_j$, $j = 1, 2, \ldots, m$.
The Lemma 2.2.1 implies that the function

$$a^2 z_m^2 - z_1^2 - \ldots - z_{m-1}^2$$

does not take non-negative values, nor does the function $a^2 z_m^2 - z_1^2 - \cdots - z_{m-1}^2 - k^2$, and it implies that every from two branches of function

$$W(z) = \sqrt{a^2 z_m^2 - z_1^2 - \cdots - z_{m-1}^2 - k^2}$$

is univalent and analytic in $T(\overset{*}{C}{}^a_+) \cup T(\overset{*}{C}{}^a_-) \equiv T(\overset{*}{C})$.

It is easy to see that the function $W(z)$ satisfies Definition 2.3.1 with $p = 1$, $a = 0$, $\alpha = 0$, $\beta = 1$ and universal constant M non-dependent on subcone $\overset{*}{C}{}^a_1$. Hence, $W(z) \in H_0(\overset{*}{C}{}^a)$, and therefore according to Proposition 2.2.3 it has boundary values in the distribution sense for $\eta \to 0$, $\eta \in \overset{*}{C}{}^a_\pm$. These boundary values are locally integrable functions and they can be easily evaluated. Since the boundary values do not depend on how η tends to zero, one can consider the following limits:

$$\lim_{\varepsilon \to +0} \sqrt{a^{(}\xi_m \pm i\varepsilon)^2 - \xi_1^2 - \cdots - \xi_{m-1}^2 - k^2}.$$

Let us choose one branch of function $W(z)$, for example that which maps $T(\overset{*}{C})$ on the upper half-plane. The boundary values of this branch will have have the following form:

$$\sqrt{a^{(}\xi_m \pm i0)^2 - \xi_1^2 - \cdots - \xi_{m-1}^2 - k^2} =$$

$$= \begin{cases} \pm\sqrt{a^2\xi_m^2 - |\xi'|^2 - k^2}, & a^2\xi_m^2 - |\xi'|^2 - k^2 > 0, \quad \xi_m > 0 \\ \mp\sqrt{a^2\xi_m^2 - \xi_1^2 - \cdots - \xi_{m-1}^2 - k^2}, & a^2\xi_m^2 - |\xi'|^2 - k^2 > 0, \quad \xi_m < 0 \\ i\sqrt{|\xi'|^2 + k^2 - a^2\xi_m^2}, & a^2\xi_m^2 - |\xi'|^2 - k^2 < 0, \end{cases}$$

which can be obtained if we will carefully observe the behavior of the modul and argument of $W(\xi_1, \ldots, \xi_{m-1}, \xi_m \pm i\varepsilon)$ for $\varepsilon \to 0$.

Further, let us note that

$$\sqrt{a^2(\xi_m + i0)^2 - |\xi'|^2 - k^2} = -\overline{\sqrt{a^2(\xi_m - i0)^2 - |\xi'|^2 - k^2}},$$

where overbar means the complex conjugation.

Turning to the function

$$\sqrt{a^2+1}\, z_m + W(z)$$

we conclude it is analytical in $T(\overset{*}{C}{}^a_+)$, and analogously, the function

$$\sqrt{a^2+1}\, z_m + \overline{W(z)}$$

is analytic in $T(\overset{*}{C}{}^a_-)$. Then one can write

$$\begin{aligned}
&\xi_m^2 + |\xi'|^2 + k^2 = \\
&= \left(\sqrt{a^2+1}\,\xi_m + \sqrt{a^2(\xi_m + i0)^2 - \xi_1^2 - \cdots \xi_{m-1}^2 - k^2}\right) \times \\
&\times \left(\sqrt{a^2+1}\,\xi_m + \overline{\sqrt{a^2(\xi_m - i0)^2 - \xi_1^2 - \cdots \xi_{m-1}^2 - k^2}}\right),
\end{aligned} \tag{5.2}$$

where each of factors admits an anlytical continuation into $T(\overset{*}{C}{}^a_+)$, $T(\overset{*}{C}{}^a_-)$ respectively. It is obvious that these analytical continuations belong to the classes $H_0(\overset{*}{C}{}^a_+)$, $H_0(\overset{*}{C}{}^a_-)$.

The equality (5.2) is the wave factorization of the Helmholtz operator. For brevity and convenience we will write it as

$$\xi_m^2 + |\xi'|^2 + k^2 =$$

$$= \left(\sqrt{a^2+1}\,\xi_m + \sqrt{a^2\xi_m^2 - |\xi'|^2 - k^2}\right)\left(\sqrt{a^2+1}\,\xi_m - \sqrt{a^2\xi_m^2 - |\xi'|^2 - k^2}\right)$$

meaning for $\sqrt{a^2\xi_m^2 - |\xi'|^2 - k^2}$ the boundary value $\sqrt{a^2(\xi_m + i0)^2 - |\xi'|^2 - k^2}$.

One can generalize this formula by considering "complex powers" of the Helmholtz operator, i.e., the operators with symbols

$$\left(\xi_m^2 + |\xi'|^2 + k^2\right)^{\alpha+i\beta}, \quad \alpha, \beta \in \mathbb{R}.$$

The wave factorization in this case will be

$$\left(\xi_m^2 + |\xi'|^2 + k^2\right)^{\alpha+i\beta} = \tag{5.3}$$

$$= \left(\sqrt{a^2+1}\,\xi_m + \sqrt{a^2\xi_m^2 - |\xi'|^2 - k^2}\right)^{\alpha+i\beta}\left(\sqrt{a^2+1}\,\xi_m - \sqrt{a^2\xi_m^2 - |\xi'|^2 - k^2}\right)^{\alpha+i\beta}$$

The analyticity of factors in $T(\overset{*}{C}{}^a_+)$, $T(\overset{*}{C}{}^a_-)$ follows from the fact that the function $\sqrt{a^2+1} + W(z)$ has no zeroes in $T(\overset{*}{C}{}^a_+)$ because for $z \in T(\overset{*}{C}{}^a_+)$ we have $\operatorname{Im} z_m > 0$, $\operatorname{Im} W(z) > 0$, and the image of this function is simply connected.

Analogous arguments are valid for the function $\sqrt{a^2+1}+\overline{W(z)}$ in domain $T(\overset{*}{C}{}^a_-)$ also.

Remark 5.1. If we consider the homogeneous symbol from class C^∞_α (α in this case is the homogeneity order) and in Definition 5.1 require the homogeneity of order æ from $A_{\neq}$ and order $\alpha -$ æ from $A_=$, then we can speak of homogeneous wave factorization.

Example 5.2. In formula (5.3) put $k=0$ and obtain the homogeneous wave factorization of symbol $|\xi|^{\alpha+i\beta}$:

$$\begin{aligned} |\xi|^{\alpha+i\beta} &= \left(\sqrt{a^2+1}\,\xi_m + \sqrt{a^2\xi_m^2 - |\xi'|^2}\right)^{\frac{\alpha+i\beta}{2}} \times \\ &\times \left(\sqrt{a^2+1}\,\xi_m - \sqrt{a^2\xi_m^2 - |\xi'|^2}\right)^{\frac{\alpha+i\beta}{2}}. \end{aligned} \tag{5.4}$$

Finally, if in formula (5.4) we let $a=0$ (in this case the $\overset{*}{C}{}^a_\pm$ "degenerate" into half-spaces $\mathbb{R}^m_\mp$), then one obtains the homogeneous factorization on variable ξ_m in the sense of Definition 4.2.2.

$$|\xi|^{\alpha+i\beta} = (\xi_m + i|\xi'|)^{\frac{\alpha+i\beta}{2}}\,(\xi_m - i|\xi'|)^{\frac{\alpha+i\beta}{2}}$$

which is well-known in Vishik-Eskin theory [92,93].

Now let $\sigma(\xi)$ be an elliptic symbol from class C^∞ for $\xi \neq 0$ and which is homogeneous of order α. According to Proposition 4.2.4 it admits the homogeneous factorization on variable ξ_m :

$$\sigma(\xi',\xi_m) = \sigma_+(\xi',\xi_m)\sigma_-(\xi',\xi_m).$$

Let us fix $\xi' = (a_1,\ldots,a_{m-1}) \equiv a'$ and consider

$$\sigma_+(a',\xi_m)\sigma_-(a',\xi_m). \tag{5.5}$$

In expression (5.5) let us make a change of variable

$$\xi_m \longmapsto \sqrt{a^2+1}\xi_m + \sqrt{a^2\xi_m^2 - |\xi'|^2}$$

for the first factor, and

$$\xi_m \longmapsto \sqrt{a^2+1}\,\xi_m - \sqrt{a^2\xi_m^2 - |\xi'|^2}$$

for the second ones, thus reducing (5.5) to the form

$$\begin{aligned} &\sigma_+\left(a', \sqrt{a^2+1}\,\xi_m + \sqrt{a^2\xi_m^2 - |\xi'|^2}\right) \times \\ &\times \sigma_-\left(a', \sqrt{a^2+1}\,\xi_m - \sqrt{a^2\xi_m^2 - |\xi'|^2}\right). \end{aligned} \tag{5.6}$$

The expression (5.6) is correct because, in spite of the fact that the functions $\left(\sqrt{a^2+1}\,\xi_m \pm \sqrt{a^2\xi_m^2 - |\xi'|^2}\right)$ take complex values, these values are in the upper and lower complex half-space, i.e., in that domain in which the functions σ_+, σ_- continued (and therefore, they are defined).

Let us denote

$$\tilde{\sigma}(a',\xi) = \sigma_+\left(a', \sqrt{a^2+1}\,\xi_m + \sqrt{a^2\xi_m^2 - |\xi'|^2}\right) \times \\ \times\sigma_-\left(a', \sqrt{a^2+1}\,\xi_m - \sqrt{a^2\xi_m^2 - |\xi'|^2}\right) \tag{5.7}$$

the symbol depending on $(m-1)$ - dimensional parameter a'.

The formula (5.7) represents the wave factorization of symbol $\tilde{\sigma}(a',\xi)$. Indeed, the analyticity of functions

$$\sigma_+\left(a', \sqrt{a^2+1}\,\xi_m + \sqrt{a^2\xi_m^2 - |\xi'|^2}\right),$$

$$\sigma_-\left(a', \sqrt{a^2+1}\,\xi_m - \sqrt{a^2\xi_m^2 - |\xi'|^2}\right)$$

in $T(\overset{*}{C}{}_+^a)$, $T(\overset{*}{C}{}_-^a)$ respectively follows from the fact that composition of analytical functions is analytical function, from the analyticity of functions σ_+, σ_- in $\mathbb{C}_+$, $\mathbb{C}_-$ ($\mathbb{C}_+$, $\mathbb{C}_-$ are upper and lower complex half-planes), and from the fact that the functions

$$\sqrt{a^2+1}\,z_m + \sqrt{a^2z_m^2 - z_1^2 - \cdots - z_{m-1}^2},$$

$$\sqrt{a^2+1}\,z_m - \sqrt{a^2z_m^2 - z_1^2 - \cdots - z_{m-1}^2},$$

are analytical as mappings $T(\overset{*}{C}{}_\pm^a) \to \mathbb{C}_\pm$ respectively.

Finally, the estimates, in which we need to satisfy the second condition of Definition 5.1, follow from homogeneity and continuity of σ_+, σ_- under arbitrary $a' \neq 0$. Although obtained under this operation symbol, $\tilde{\sigma}(a',\xi)$ does not belong generally speaking to the class C_α^∞, but it satisfies the inequality (5.1), and therefore pseudodifferential operator with such symbol boundedly acts from the space $H_s(\mathbb{R}^m)$ into the space $H_{s-\alpha}(\mathbb{R}^m)$ according to Proposition 4.1.1.

Further, we will consider a special type of symbols, namely symbols of multidimensional singular integral operators of Calderon-Zygmund.

Let $A(\xi)$ be homogeneous of order 0 elliptic symbol from class $C^\infty(\mathbb{R}^m \setminus \{0\})$ (and, hence satisfying the condition (5.1) with power $\alpha = 0$). Under fixed ξ' it can be represented as a convergent series

$$A(\xi) = \sum_{k=-\infty}^{+\infty} c_k \left(\frac{\xi_m - i|\xi'|}{\xi_m + i|\xi'|}\right)^k,$$

because the function system $\left\{\left(\frac{\xi_m - i|\xi'|}{\xi_m + i|\xi'|}\right)^k\right\}_{-\infty}^{+\infty}$ generates the basis in space $L_2(\mathbb{R})$. Let us assume that $A(\xi)$ admits factorization on variable ξ_m in the sense of Definition 4.2.2

$$A(\xi) = A_+(\xi)A_-(\xi)$$

in the class of convergent series

$$A_+(\xi) = \sum_{k=0}^{+\infty} a_k \left(\frac{\xi_m - i|\xi'|}{\xi_m + i|\xi'|}\right)^k, \tag{5.8}$$

$$A_-(\xi) = \sum_{k=-\infty}^{-1} b_k \left(\frac{\xi_m - i|\xi'|}{\xi_m + i|\xi'|}\right)^k . \tag{5.9}$$

It is obvious that $A(\xi)$, $A_+(\xi)$, $A_-(\xi)$ are homogeneous of order zero functions, and therefore the factorization index on ξ_m - variable is equal to 0. Our assumption is equivalent to saying that $A(0,\ldots,0,+1) = A(0,0,\ldots,0,-1)$, and the increment of the argument of function $A(\xi)$ along any big half-circumference from unit sphere S^{m-1} which joins the points $(0,\ldots,0,-1)$ and $(0,\ldots,0,+1)$ (in case $m=2$ right and left half-circumference) is equal to zero. Actually, in this case the factorization on ξ_m exists according to Proposition 4.2.4, and A_+, A_- as continuous functions (under fixed ξ')expand into convergent Fourier series on positive and negative powers respectively. Multiplying (5.8) and (5.9) we obtain the collection of relations

$$\begin{aligned} c_0 &= a_1 b_{-1} + a_2 b_{-2} + \cdots + a_n b_{-n} + \cdots \\ c_1 &= a_2 b_{-1} + a_3 b_{-2} + \cdots + a_n b_{-(n-1)} + \cdots \\ c_{-1} &= a_1 b_{-2} + a_2 b_{-3} + \cdots + a_n b_{-(n+1)} + \cdots \\ &\ldots\ldots\ldots\ldots\ldots\ldots\ldots\ldots\ldots , \end{aligned} \tag{5.10}$$

and every numerical series of infinite system (5.10) is convergent.

Let us consider now the collection $\mathcal{B}$ of homogeneous of order zero functions which can be represented by convergent series

$$\overset{\vee}{A}(\xi) = \sum_{k=-\infty}^{+\infty} c_k \left(\frac{\sqrt{a^2+1} - \sqrt{a^2\xi_m^2 - |\xi'|^2}}{\sqrt{a^2+1} + \sqrt{a^2\xi_m^2 - |\xi'|^2}}\right)^k ,$$

Let us denote

$$A_{\neq}(\xi) = \sum_{k=0}^{+\infty} a_k \left(\frac{\sqrt{a^2+1} - \sqrt{a^2\xi_m^2 - |\xi'|^2}}{\sqrt{a^2+1} + \sqrt{a^2\xi_m^2 - |\xi'|^2}}\right)^k , \tag{5.11}$$

$$A_{=}(\xi) = \sum_{k=-\infty}^{-1} b_k \left(\frac{\sqrt{a^2+1} - \sqrt{a^2\xi_m^2 - |\xi'|^2}}{\sqrt{a^2+1} + \sqrt{a^2\xi_m^2 - |\xi'|^2}}\right)^k . \tag{5.12}$$

From previous reasonings easily follows

Theorem 5.1. *Let $\overset{\vee}{A}(\xi) \in \mathcal{B}$. If there exist sequences of numbers $\{a_k\}_{k=0}^{\infty}$, $\{b_k\}_{k=-\infty}^{-1}$ such that the series (5.11), (5.12) are convergent and relations (5.10) are satisfied, then $\overset{\vee}{A}(\xi)$ admits the homogeneous wave factorization with index zero.*

It seems at first the Theorem 5.1 is difficult to apply in any practical manner, but in the case of finit sums it can be done simply.

Example 5.3. Let $a = 1$,

$$\overset{\vee}{A}(\xi) = 3\ell^{-1}(\xi) + 7 + 2\ell(\xi),$$

where

$$\ell(\xi) = \frac{\sqrt{2}\,\xi_m - \sqrt{\xi_m^2 - |\xi'|^2}}{\sqrt{2}\,\xi_m + \sqrt{\xi_m^2 - |\xi'|^2}}.$$

The symbol $\overset{\vee}{A}(\xi)$ is elliptic because $\overset{\vee}{A}(\xi) \neq 0$, $\forall \xi \neq 0$. Indeed, $\overset{\vee}{A}(\xi) = 0$ if only $\ell(\xi) = -3$, or $1/2$, but the function $\ell(\xi)$ does not take these values. Let us verify this, for example, for -3.

If $\ell(\xi) = -3$, then

$$\sqrt{2}\,\xi_m - \sqrt{\xi_m^2 - |\xi'|^2} = -3\sqrt{2}\xi_m - 3\sqrt{\xi_m^2 - |\xi'|^2},$$

$$28\xi_m^2 + 4|\xi'| = 0.$$

The last is possible only if $\xi_m = 0$, $|\xi'| = 0$, i.e. $\xi = 0$.

The value $-1/2$ can be verified analogously.

Seek $A_{\neq}$, $A_{=}$ in the form

$$A_{\neq}(\xi) = a_0 + a_1\ell(\xi),$$

$$A_{=}(\xi) = b_0 + b_{-1}\ell(\xi).$$

The relations (5.10) give the system

$$\begin{cases} a_0b_0 + a_1b_{-1} & = 7 \\ a_1b_0 & = 2 \\ a_0b_{-1} & = 3 \end{cases}$$

from which we obtain $b_0 = 2a_1^{-1}$, $b_{-1} = 3a_0^{-1}$.

Substituting these values into the first equation we have

$$2\frac{a_0}{a_1} + 3\frac{a_1}{a_0} = 7.$$

Denoting $a_0a_1^{-1} = k$, we obtain the equation

$$2k + \frac{3}{k} = 7,$$

which has two roots $k_1 = 1/2$, $k_2 = 3$, as a result two variants of wave factorization appear

$$\overset{\vee}{A}(\xi) = (1 + 2\ell(\xi))(1 + 3\ell^{-1}(\xi)),$$

$$\overset{\vee}{A}(\xi) = (3 + \ell(\xi))(2 + \ell^{-1}(\xi))$$

(note carefully that wave factorization is obtained up to an arbitrary constant), but the first variant is not suitable because the function $1 + 2\ell(z)$ has zeroes in $T(\overset{*}{C}{}_+^1)$, and the function $1+3\ell^{-1}(z)$ in $T(\overset{*}{C}{}_-^1)$. Indeed, let us consider, for example, the equation

$$1 + 2\ell(z) = 0, \quad z \in T(C_+^1)$$

(in the case $a = 1$ we have $\overset{*}{C}{}_+^1 = C_+^1$).

Then

$$0 = 1 + 2\ell(z) =$$

$$= 1 + 2\frac{\sqrt{2}\, z_m + \sqrt{z_m^2 - z_1^2 - \cdots - z_{m-1}^2}}{\sqrt{2}\, z_m + \sqrt{z_m^2 - z_1^2 - \cdots - z_{m-1}^2}} =$$

$$= \frac{3\sqrt{2}\, z_m + 3\mathrm{Re}\sqrt{z_m^2 - z_1^2 - \cdots - z_{m-1}^2} - i\mathrm{Im}\sqrt{z_m^2 - z_1^2 - \cdots - z_{m-1}^2}}{\sqrt{2}\, z_m + \sqrt{z_m^2 - z_1^2 - \cdots - z_{m-1}^2}}.$$

Since the denominator of the last fraction has no zeroes in $T(C_+^1)$, we seek such points from $T(C_+^1)$ in which

$$\begin{aligned} &3\sqrt{2}\, z_m + 3\mathrm{Re}\sqrt{z_m^2 - z_1^2 - \cdots - z_{m-1}^2} - \\ &-i\mathrm{Im}\sqrt{z_m^2 - z_1^2 - \cdots - z_{m-1}^2} = 0. \end{aligned} \tag{5.13}$$

Let

$$x_m = 0,\ y_1 = \cdots = y_{m-1} = 0.$$

The equation (5.13) will take the form

$$\begin{aligned} &3\sqrt{2}\, iy_m + 3\mathrm{Re}\sqrt{-y_m^2 - x_1^2 - \cdots - x_{m-1}^2} - \\ &-i\mathrm{Im}\sqrt{-y_m^2 - x_1^2 - \cdots - x_{m-1}^2} = 0. \end{aligned} \tag{5.14}$$

(5.14) implies that

$$3\sqrt{2}\, iy_m - i\sqrt{y_m^2 + x_1^2 + \cdots + x_{n-1}^2} = 0,$$

and it gives

$$y_m = \frac{1}{\sqrt{17}}|x'|, \quad y_m > 0. \tag{5.15}$$

Hence, we have demonstrated that the points $z = (z_1, \ldots, z_m)$ for which their coordinates satisfy the condition (5.15) (they obviously belong to $T(C_+^1)$) are solutions of equation

$$1 + 2\ell(z) = 0.$$

Of course one can consider more complicated examples of wave factorization for functions of type

$$\overset{\vee}{A}(\xi) = \sum_{k=-n}^{n} c_k \ell^k(\xi).$$

Now we will consider the question of uniqueness of wave factorization.

Theorem 5.2. *If an elliptic symbol $A(\xi)$ from class C_α admits wave factorization with respect to cone C_+^a, then it is unique up to a multiplicative constant.*

Proof. Let us assume wave factorization

$$A(\xi) = A_{\neq}(\xi) A_{=}(\xi)$$

with index æ and another wave factorization

$$A(\xi) = B_{\neq}(\xi)B_{=}(\xi)$$

with index γ. Let us write

$$A_{\neq}(\xi)B_{\neq}^{-1}(\xi) = A_{=}^{-1}(\xi)B_{=}(\xi).$$

Then $A_{\neq}(z)B_{\neq}^{-1}(z)$ is analytic in $T(\overset{*}{C}{}_{+}^{a})$ with estimate

$$|A_{\neq}(z)B_{=}^{-1}(z)| \le c(1+|z|)^{æ-\gamma},$$

and $A_{=}^{-1}(z)B_{=}(z)$ is analytic in $T(\overset{*}{C}{}_{-}^{a})$ with estimate

$$|A_{=}^{-1}(z)B_{=}(z)| \le c(1+|z|)^{æ-\alpha+\alpha-\gamma} = c(1+|z|)^{æ-\gamma},$$

i.e., $A_{\neq}B_{\neq}^{-1} \in H_0(\overset{*}{C}{}_{+}^{a})$, $A_{=}^{-1}B_{=} \in H_0(\overset{*}{C}{}_{-}^{a})$.

Then according to corollary 2.2.2 the function

$$f(z) = \begin{cases} A_{\neq}(z)B_{\neq}^{-1}(z), & z \in T(\overset{*}{C}{}_{+}^{a}) \\ A_{=}^{-1}(z)B_{=}(z), & z \in T(\overset{*}{C}{}_{-}^{a}) \end{cases}$$

belongs to $H_0(C)$, where C is the convex hull of $\overset{*}{C}{}_{+}^{a} \cup \overset{*}{C}{}_{-}^{a}$, which obviously coincides with $\mathbb{R}^m$. Furthermore this function $f(z)$ has a power order at infinity and has no zeroes in $T(\overset{*}{C}{}_{+}^{a}) \cup T(\overset{*}{C}{}_{-}^{a})$. By Liouville theorem (Corollary 2.3.3) $f(z)$ must be a polynomial but every polynomial will have zeroes either in $\mathbb{R}^m$ or in $T(\overset{*}{C}{}_{+}^{a}) \cup T(\overset{*}{C}{}_{-}^{a})$. We have only unique possibility, namely $f(z) = \text{const}$ and hence

$$A_{\neq}(\xi) = cB_{\neq}(\xi), \quad A_{=}(\xi) = c^{-1}B_{=}(\xi).$$

It is clear that, given a "normalization condition" of (4.2.9) type, then wave factorization of symbol $A(\xi)$ will be unique. ■

Remark 5.2. Such factorization in cones for ellptic symbols $A(\xi)$ was used earlier in papers [283,284] for solution of certain multidimensional linear conjugation problems (see Appendix 1 also), and under very strong restrictions on factorization elements, it turns out that the condition

$$\text{supp}\, F_{\xi\to x}^{-1}(\ln(A(\xi)) \subseteq (C_{+}^{a} \cup C_{-}^{a}) \tag{5.16}$$

(see Appendix 1, Theorem A1.3) is necessary and sufficient for existence of wave factorization. In our situation condition (5.16) is not necessary. Indeed, let C_{+}^{1} be a future light cone ($a \equiv 1$). Let us consider symbol

$$A'(\xi) = \frac{\sqrt{2}\,\xi_m - \sqrt{\xi_m^2 - |\xi'|^2}}{\sqrt{2}\,\xi_m + \sqrt{\xi_m^2 - |\xi'|^2}}. \tag{5.17}$$

Evidently symbol $A'(\xi)$ is homogeneous of order 0. If for $A'(\xi)$ the condition (5.16) holds, then according to theorem A1.3 (Appendix 1; let us note that infinite differentiability of the symbol in this theorem is not essential) we have wave factorization with index 0

$$A'(\xi) = A'_{\neq}(\xi)A'_{=}(\xi). \tag{5.18}$$

On the other hand $A'(\xi)$ admits natural wave factorization (5.17). Comparing (5.17) and (5.18) we have equality

$$A'_{\neq}(\xi)\left(\sqrt{2}\,\xi_m + \sqrt{\xi_m^2 - |\xi'|^2}\right) = \left(A'_{=}\right)^{-1}(\xi)(\sqrt{2}\,\xi_m - \sqrt{\xi_m^2 - |\xi'|^2}).$$

Now from " edge of the wedge" theorem and Liouville's theorem (Proposition 2.3.3 and Corollary 2.3.3) follows factorization elements, having no zeroes, must be polynomials of order no more than 1. The unique possibility, constant, is not suitable because in this case functions of order 0 and order 1 are homogeneously equal.

The index of wave factorization generally speaking does not coincide with the factorization index on ξ_m - variable. (They do coincide in the situation considered above). The example suggested below describes the situation when the index of wave factorization is 0, but the factorization index on ξ_m - variable generally speaking is non-vanishing.

Let $m \geq 3$, $\omega(\xi)$ be a real homogeneous function of order 0, and for $\exp\omega(\xi)$ condition (5.16) holds; in addition, let $\omega(\xi)$ be continuous on S^{m-1}. It is obvious that the class of such ω is non-empty. Evidently, the condition (5.16) will be valid also for $\exp(i\omega)$. The factorization index on ξ_m - variable for such a symbol is equal to (see [92,93])

$$\frac{\omega(0,+1) - \omega(0,-1)}{2\pi} + m' ,$$

where m' is an integer, but the difference $\omega(0,+1) - \omega(0,-1)$ is not necessary a multiple of 2π.

Finally, note that selection of the term "wave factorization" is stipulated by properties of so-called causal functions and their relation with wave equation [30,190,286].

6. Diffraction on a quadrant

Diffraction theory, one of the most important areas of mathematical physics, has recently been under intensive development. Integral or more generally pseudodifferential equations take an exceptional place in diffraction theory; as a rule diffraction problems reduce to one or another of them [69,123,57,118]. This reduction to boundary equations is achieved by different methods, depending on the particular equation type, and the investigations in many cases lead to existence and uniqueness theorems. One of the methods used for solving the boundary equations thus attained is the Wiener-Hopf method, or the factorization method, which has been successfully applied to some diffraction problems [166-168,175,240,256,257]. We suggested here a multidimensional generalization of this method, and with its help we will study pseudodifferential equations arising from a diffraction problem on a quadrant, obtained in [168]. The solution in the simplest case of this problem can be written in explicit form and is more appealing than the formula found in [168].

From a mathematical point of view, the problem that we will consider can be formulated in the following way [168]: Find the function $u \in H_1(\mathbb{R}^3)$ satisfying the Helmholtz equation

$$(\Delta + k^2)u(x) = 0, \quad x \in \mathbb{R}^3 \setminus \Gamma^+, \tag{6.1}$$

where Δ is the Laplacian

$$\Delta = \frac{\partial^2}{\partial x_1^2} + \frac{\partial^2}{\partial x_2^2} + \frac{\partial^2}{\partial x_3^2},$$

$\Gamma^+ = \{x \in \mathbb{R}^3 : \; x_3 = 0, \; x_2 > |x_1|\}$, with Dirichlet boundary condition

$$u(x') = g(x'), \quad x' \in \Gamma^+, \tag{6.2}$$

or Neumann condition

$$\frac{\partial u}{\partial x_3}(x') = h(x'), \quad x' \in \Gamma^+ \; (x_3 = \pm 0) \tag{6.3}$$

and g, h are arbitrary given functions, $g \in H_{1/2}(\Gamma^+)$, $h \in H_{-1/2}(\Gamma^+)$, $k \in \mathbb{C}$, $k = k_1 + ik_2$, $k_2 > 0$, $x' = (x_1, x_2)$.

Let B be pseudodifferential operator with symbol $\sigma(\xi') = (\xi_1^2 + \xi_2^2 - k^2)^{1/2}$, χ_M be multiplication operator on the characteristic function of set M, $H_s(M)$ be subspace in $H_s(\mathbb{R}^2)$ consisting of functions $u(x')$ with support in $\overline{M}$.

E.Meister and F.-O.Speck proved the following theorem on equivalense and representation of solution of the Dirichet problem (6.1), (6.2) (we give it in our notation).

Theorem 6.1. *The function u is a solution of the Dirichlet problem (6.1), (6.2) if and only if the following equalities are valid:*

$$u(x) = Ku_0(x) =$$

$$= F^{-1}_{\xi' \to x'} \exp\left(-|x_3|\sigma(\xi')\right) \tilde{u}_0(\xi'),$$

$$u_0 = B^{-1}w_+, \quad w_+ \in H_{-1/2}(\Gamma^+),$$

and w_+ is a solution of the Wiener-Hopf equation

$$\chi_{\Gamma^+} B^{-1} w_+ = g. \tag{6.4}$$

Thus, to solve the diffraction problem (6.1), (6.2) on a quadrant (our main attention will be given to the Dirichlet problem) one needs to solve the equation (6.4).

We give the key result of paper [168], also related to solution of the Dirichlet problem (6.1), (6.2), to compare with our result. Note that in [168] the diffraction problem was considered on the screen $\sum_1 = \{x \in \mathbb{R}^3 : \ x_1 > 0, x_2 > 0, x_3 = 0\}$ instead of Γ^+. The methods of paper [168] are based on a one-dimensional factorization technique.

Let $t_{\pm(1)}(\xi') = \left(\xi_1 \pm i(\xi_2^2 - k^2)^{1/2}\right)^{1/2}$, $B_{\pm(1)}$ be pseudodifferential operators with symbols $t_{\pm(1)}(\xi')$ respectively, and analogously, $B_{\pm(2)}$ be pseudodifferential operators with symbols $t_{\pm(2)}(\xi') = \left(\xi_2 \pm i(\xi_1^2 - k^2)^{1/2}\right)^{1/2}$, $\sum_{14} = \{x \in \mathbb{R}^3 : \ x_1 > 0, x_2 \in \mathbb{R}, x_3 = 0\}$, $\sum_{12} = \{x \in \mathbb{R}^3 : \ x_1 \in \mathbb{R}, x_2 > 0, x_3 = 0\}$.

We denote Π_1, Π_2 the projectors

$$\Pi_1 = B_{-(1)}^{-1} \chi_{\sum_{14}} B_{-(1)},$$

$$\Pi_2 = B_{-(2)}^{-1} \chi_{\sum_{12}} B_{-(2)};$$

$R(\Pi_1)$, $R(\Pi_2)$ denote images of operators Π_1, Π_2.

Theorem 6.2. *Let $k = i$. The solution of Dirichlet problem (6.1), (6.2) on screen $\sum_1$ can be represented in the form*

$$u = K \Pi \ell g$$

for arbitrary function $g \in H^{1/2}(\sum_1)$, where ℓg is a arbitrary continuation of g into $H_{1/2}(\mathbb{R}^2)$, $\Pi = \Pi_1 \wedge \Pi_2$ is the infimum from Π_1 and Π_2, i.e., Π is an orthogonal projector on $R = R(\Pi_1) \wedge R(\Pi_2)$. Π is expressed by the formula

$$\Pi = \lim_{n \to \infty} \Pi_1 (\Pi_2 \Pi_1)^n = \lim_{n \to \infty} \Pi_2 (\Pi_1 \Pi_2)^n,$$

and the limits one means in the sense of strong $H_{1/2}$ - convergence.

In papers [166,167] one can find a survey of some results related to this and similar problems.

Let $k = ik_2$, i.e., $k_1 = 0$, for the remainder of this chapter we will assume this condition holds.

The symbol of operator B^{-1} is equal to

$$\sigma^{-1}(\xi') = (\xi_2^2 - \xi_1^2 - k^2)^{-1/2}.$$

For brevity, let us denote

$$\sigma_{\neq}(\xi') = \left(\sqrt{2}\xi_2 + (\xi_2^2 - \xi_1^2 + k^2)^{1/2}\right)^{-1/2},$$

$$\sigma_{=}(\xi') = \left(\sqrt{2}\xi_2 - (\xi_2^2 - \xi_1^2 + k^2)^{1/2}\right)^{-1/2},$$

meaning as earlier for $(\xi_2^2 - \xi_1^2 + k^2)^{1/2}$ the boundary value of function $(z_2^2 - z_1^2 + k^2)^{1/2}$ for $z = (z_1, z_2) \in T(\Gamma^+)$.

Then

$$\sigma^{-1}(\xi') = \sigma_{\neq}(\xi')\sigma_{=}(\xi') \tag{6.5}$$

is a wave factorization of symbol $\sigma^{-1}(\xi')$ with respect to cone Γ^+ for radial tube domains $T(\Gamma^+)$, $T(\Gamma^-)$, $\Gamma^- = -\Gamma^+$.

Let us begin to solve the equation (6.4). Let g_1 be arbitrary continuation of g from Γ^+ on $\mathbb{R}^2$, $g_1 \in H_{1/2}(\mathbb{R}^2)$. Let

$$w_- = g_1 - B^{-1}w_+,$$

rewrite the equation (6.4) in the form

$$B^{-1}w_+ + w_- = g_1, \tag{6.6}$$

and apply to (6.6) the Fourier transform:

$$\sigma^{-1}(\xi')\tilde{w}_+(\xi') + \tilde{w}_-(\xi') = \tilde{g}_1(\xi'). \tag{6.7}$$

Taking into account the wave factorization (6.5) let us denote

$$\tilde{v}_+(\xi') = \sigma_{\neq}(\xi')\tilde{w}_+(\xi'),$$

$$\tilde{v}_-(\xi') = \sigma_{=}^{-1}(\xi')\tilde{w}_-(\xi'),$$

$$\tilde{g}_2(\xi') = \sigma_{=}^{-1}(\xi')\tilde{g}_1(\xi').$$

We write (6.7) in the form

$$\tilde{v}_+(\xi') + \tilde{v}_-(\xi') = \tilde{g}_2(\xi'). \tag{6.8}$$

Lemma 6.1. $\tilde{v}_+ \in A_2(\mathbb{R}^2)$, $\tilde{v}_- \in B_2(\mathbb{R}^2)$, $\tilde{g}_2 \in L_2(\mathbb{R}^2)$. [2])

Proof. First we have $\tilde{v}_\pm, \tilde{g}_2 \in L_2(\mathbb{R}^2)$ because the operators with symbols $\sigma_{\neq}$, $\sigma_{=}^{-1}$ have orders $-1/2$ and $1/2$ respectively, and they boundedly act from spaces $H_{-1/2}(\mathbb{R}^2)$ and $H_{1/2}(\mathbb{R}^2)$ into space $H_0(\mathbb{R}^2) = L_2(\mathbb{R}^2)$ according to Proposition 4.1.1. Further, since $\sigma_{\neq}$ and $\tilde{w}_+$ admit an analytical continuation into $T(\Gamma^+)$, then the $\tilde{v}_+$ has the same property. Hence, $\tilde{v}_+ \in A_2(\mathbb{R}^2)$ because

$$\int\limits_{\mathbb{R}^2} |\tilde{v}_+(\xi' + i\eta')|^2 \, d\xi' \leq$$

$$\leq \text{const} \int\limits_{\mathbb{R}^2} |\tilde{w}_+(\xi' + i\eta')|^2 \,(1 + |\xi'|)^{-1} d\xi' \leq \text{const}, \quad \eta' \in \Gamma^+,$$

by virtue of properties of the factorization element $\sigma_{\neq}$; the last inequality follows from the fact that $\tilde{w}_+ \in \tilde{H}^+_{-1/2}(\mathbb{R}^2)$.

Second, $\sigma_{=}^{-1}$ admits an analytical continuation into $T(\Gamma^-)$ from class $H_0(\Gamma)$, and hence the inverse Fourier transform of this function is a distribution a for which its support is belongs to $\overline{\Gamma^-}$ according to Proposition 2.3.2. Since supp $w_- \subset$

[2]) see Appendix 1

$\mathbb{R}^2 \setminus \Gamma^+$, $v_- = a * w$ in a distribution sense, and "$*$" denotes the convolution, then it is easily verified that $\operatorname{supp} v_- \subset \mathbb{R}^2 \setminus \Gamma^+$. Actually, if w_- is a basic function, i.e., it is infinitely differentiable with compact support in $\mathbb{R}^2 \setminus \Gamma^+$, then the convolution $a * w_-$ is defined by formula

$$a * w_- = \left(a(y), \overline{w_-(x-y)}\right),$$

where $w_-(x-y)$ is considered as a function on y (x is fixed), and notation $a(y)$ means that functional a acts on the basic function on the y - variable of the basic function.]

Let $x \in \Gamma^+$. If $y \notin \Gamma^-$, then since $\operatorname{supp} a \subset \Gamma^-$ we have $(a * w_-)(x) = 0$. If $y \in \Gamma^-$ then $x - y \in \Gamma^+$, and $(a * w_-)(x) = 0$ by virtue of $\operatorname{supp} w_- \in \mathbb{R}^2 \setminus \Gamma^+$. The last means that $\tilde{v}_- \in L_2(\mathbb{R}^2) \ominus A_2(\mathbb{R}^2) = B_2(\mathbb{R}^2)$.

Extention to the general case is realized by the usual approximation method. ■

We introduce the integral operator G_2 which is defined on functions $u \in S(\mathbb{R}^2)$ by the formula

$$(G_2 u)(x') = \frac{1}{\pi^2} \lim_{\tau \to +0} \int_{\mathbb{R}^2} \frac{u(y')dy'}{(x_1 - y_1)^2 - (x_2 - y_2 + i\tau)^2} - u(x'), \quad x' \in \mathbb{R}^2,$$

(the limit is meaning in the sense of L_2 - convergence) and by two projectors

$$C_2^+ = \frac{1}{2}(I + G_2),$$

$$C_2^- = \frac{1}{2}(I - G_2).$$

By Lemma $A1.1$ from Appendix 1, G_2^+ is orthogonal projector on subspace $A_2(\mathbb{R}^2)$, G_2^- on $B_2(\mathbb{R}^2)$.

Now we are ready to prove the main result on solution of the Wiener-Hopf equation (6.4).

Theorem 6.3. *The Wiener-Hopf equation (6.4) under arbitrary right-hand side $g \in H_{1/2}(\Gamma^+)$ has a unique solution $w_+ \in H_{-1/2}(\Gamma^+)$ which can be written as formula*

$$w_+ = F^{-1} \sigma_{\neq}^{-1} G_2^+ \sigma_{=}^{-1} F g_1, \tag{6.9}$$

where g_1 is an arbitrary continuation of g on $\mathbb{R}^2$, $g_1 \in H_{1/2}(\mathbb{R}^2)$.

Proof. Above we reduced the Wiener-Hopf equation (6.4) to a jump problem (6.8) which is uniquely solvable by virtue of Lemma $A1.2$ from Appendix 1, thus

$$\tilde{v}_\pm = G_2^\pm \tilde{g}_2,$$

and we obtain

$$\tilde{w}_+ = \sigma_{\neq}^{-1} G_2^+ \sigma_{=}^{-1} \tilde{g}_1. \tag{6.10}$$

Applying the inverse Fourier transform we obtain (6.9). ■

It is necessary to note that result (6.10) does not depend on we choice of continuation g_1. Indeed, let g_1' be another continuation from $H_{1/2}(\Gamma^+)$ on $H_{1/2}(\mathbb{R}^2)$,

$$\tilde{w}'_+ = \sigma_{\neq}^{-1} G_2^+ \sigma_{=}^{-1} \tilde{g}'_1 .$$

Then

$$\tilde{w}_+ - \tilde{w}'_+ = \sigma_{\neq}^{-1} G_2^+ \sigma_{=}^{-1} (\tilde{g}_1 - \tilde{g}'_1) .$$

Since g_1 and g_1' coincide on Γ^+, then $\operatorname{supp}(g_1 - g_1') \subset \mathbb{R}^2 \setminus \Gamma^+$.

Further, analogous to the proof of Lemma 6.1 one determines that $\sigma_{=}^{-1}(\tilde{g}_1 - \tilde{g}'_1) \in \tilde{H}_0^-(\mathbb{R}^m) = \tilde{L}_2(\mathbb{R}^2 \setminus \Gamma^+)$, and since G_2^+ is a projector on the complementary subspace $\tilde{L}_2(\Gamma^+) = A_2(\mathbb{R}^2)$, then

$$G_2^+ \sigma_{=}^{-1} (g_1 - g_1') = 0,$$

i.e., $\tilde{w}_+ = \tilde{w}'_+$.

It seems, the formula (6.9) for solution of equation (6.4) is more successful with respect to that formula from Theorem 6.2 because for all elements contained in (6.9) one has explicit description.

According to Theorems 6.1 and 6.3 one deduces

Theorem 6.4. *The Dirichlet problem (6.1), (6.2) for arbitrary $g \in H_{1/2}(\Gamma^+)$ has a unique solution $u \in H_1(\mathbb{R}^3)$ which is written in the form*

$$\begin{aligned} u(x) = F^{-1}_{\xi' \to x'} \exp\left(-|x_3| \left(\xi_1^2 + \xi_2^2 - k^2\right)^{1/2}\right) \times \\ \times F^{-1}_{x' \to \xi'} B^{-1} F^{-1} \sigma_{\neq}^{-1} G_2^+ \sigma_{=}^{-1} F g. \end{aligned} \tag{6.11}$$

In formula (6.11) x_3 is fixed. The value of function u at point x is added from $x' \left(F^{-1}_{\xi' \to x'}\right)$ and x_3.

We now proceed to consideration of the Neumann problem (6.1), (6.3). In paper [168] the problem was also reduced to some Wiener-Hopf equation of (6.4) type. We will briefly discuss Neumann problem and plan ways for its solution. We refer to the some results of paper [168].

Theorem 6.5. *The function u is solution of the Neumann problem (6.1), (6.3) if and only if*

$$u(x) = -\operatorname{sgn} x_3 K u_+(x), \tag{6.12}$$

where $u_+ \in H_{1/2}(\Gamma^+)$ is the solution of the Wiener-Hopf equation

$$\chi_{\Gamma^+} B_k u_+ = h, \tag{6.13}$$

B_k is a pseudodifferential operator with symbol $\left(\xi_1^2 + \xi_2^2 - k^2\right)^{1/2}$:

$$B_k = F^{-1} \left(\xi_1^2 + \xi_2^2 - k^2\right)^{1/2} F.$$

We introduce the operators

$$\tilde{\Pi}_1 = B^{-1}_{+(1)} \chi_{14} B_{+(1)},$$

$$\tilde{\Pi}_2 = B_{+(2)}^{-1}\chi_{12}B_{+(2)},$$

$$\tilde{\Pi} = \tilde{\Pi}_1 \wedge \tilde{\Pi}_2.$$

Theorem 6.6. *The solution of the Neumann problem (6.1), (6.3) is given by formula*

$$u = -\operatorname{sgn} x_3 \tilde{W}_0^{-1}\tilde{\Pi} B_i^{-1}\ell h,$$

where the operator

$$\tilde{W}_0 = \tilde{\Pi} B_k B_i^{-1}\Big|_{R(\tilde{\Pi})}$$

is iverted by a Neumann series and is absent in case $k = 1$, ℓh is arbitrary continuation of $h \in H_{-1/2}(\sum_1)$ onto $H_{-1/2}(\mathbb{R}^2)$. (Recall that in [168] the Neumann problem was considered on screen $\sum_1$).

In case $k = i$, for solution of equation (6.13) we suggest applying the wave factorization method as above. The symbol of operator B_i has the form

$$\left(\xi_1^2 + \xi_2^2 + 1\right)^{1/2}$$

and it factorizes in the following way

$$\left(\xi_1^2 + \xi_2^2 + 1\right)^{1/2} =$$

$$= \left(\sqrt{2}\xi_2 + \left(\xi_2^2 - \xi_1^2 - 1\right)^{1/2}\right)^{1/2}\left(\sqrt{2}\xi_2 - \left(\xi_2^2 - \xi_1^2 - 1\right)^{1/2}\right)^{1/2},$$

which that the scheme of proof for Theorem 6.3 completely repeats. The resulting solution u_+ substitutes into formula (6.12), and the Neumann problem (6.1), (6.3) is solved.

In conclusion we note that the Wiener-Hopf equation (6.4) and (6.13), in the case of complex wave numbers $k = k_1 + ik_2$, $k_2 > 0$, can be solved with the help of Neumann series (as suggested in [168]) taking into account the "sectoriality" property [240] for symbol $\Phi_0(\xi') = \left(\xi_1^2 + \xi_2^2 + 1\right)^{1/2}\left(\xi_1^2 + \xi_2^2 - k^2\right)^{-1/2}$.

7. The problem of indentation of a wedge-shaped punch

One important class of problems considered in elasticity theory is the class of so-called contact problems. Its investigation was begun at the end of 19th century; it significantly advanced [67,106,141,142,143,178,202,212] by application of different methods and results of differential and integral equations theory. In recent years methods of integral equations (potential theory) have been greatly extended because, as a rule, contact problems reduce to equations of this type [192,193]. But solution of these integral equations meets with serious mathematical difficulties. A large number of papers is devoted to the study of special types of such equations [106,178] when a punch has fixed form (for example, the punch is circular, elliptical, or wedge-shaped, etc.), and in these papers they develop asymptotic methods of solution.

Here the case of a wedge-shaped punch will be considered. This problem has already attracted the attention of researchers [7,211], and in numerous cases it has been solved. For example, in [23] a method for solving of corresponding integral equations was described, based on expansion of the kernel in series followed by application of a Mellin transform. As a result, the given integral equation is reduced to an infinite system of linear algebraic equations, which suggests applcation of a reduction method in a sequence space with some weight function. We will look in more detail at this integral equation by wave factorization method which is a multidimensional generalization of the well known Wiener-Hopf method.

In case of static contact without friction of wedge-shaped punch in structure with elastic half-space the problem (from a mathematical point of view) is reduced to solving of the following integral equation

$$\iint\limits_{\Omega} k(x-\xi, y-\eta)q(\xi,\eta)d\xi d\eta = C_1 f(x,y), \tag{7.1}$$

where $f(x,y)$ is a given function in a two-dimensional domain

$$\Omega = \{(x,y) \in \mathbb{R}^2 : \ y > a|x|, \ a > 0\},$$

$q(\xi,\eta)$ is an unknown function, C_1 is defined by elastic constants of the half-space, and the kernel of integral equation (7.1) is defined by the inverse Fourier transform

$$k(x,y) = \frac{1}{4\pi^2} \iint\limits_{\mathbb{R}^2} (\xi^2+\eta^2)^{-1/2} e^{-(x\xi+y\eta)} d\xi d\eta.$$

Here, in comparison with the equation that is written out in [23], we will introduce for convenience some unessential modifications in the definition of domain Ω. For the domain considered the wedge-shaped punch may be only convex although at the same time this restriction in [23] is absent. But it will be evident below that in the non-convex case the problem can be solved by an analogous method.

The equation (7.1) generally speaking belongs to a class of pseudodifferential equations which we will consider in Chapter 8. Of course it can be considered from the integral equations point of view, but for this one needs to pay attention to defining the integral contained in this equation.

The equation (7.1) will be considered in the context of the theory of pseudodifferential eqations in H_s - space scale of S.L.Sobolev – L.N.Slobodetskii. The scheme suggested here will be developed and generalized in Chapter 8.

Let us denote by K the following operator generated by the kernel of equation (7.1):

$$(Ku)(x,y) = \iint\limits_{\mathbb{R}^2} k(x-x', y-y')u(x',y')dx'dy'. \tag{7.2}$$

According to the definition of pseudodifferential operator (section 4) the operator K can be considered as a pseudodifferential operator with symbol $(\xi^2+\eta^2)^{-1/2}$ and (7.2) is an integral representation of operator K.

Let us introduce into consideration the operator K_ε corresponding to symbol

$$\sigma(\xi,\eta,\varepsilon)=(\xi^2+\eta^2+\varepsilon^2)^{-1/2}.$$

We write the equation (7.1) in operator form

$$P_+Kq=f, \tag{7.3}$$

where P_+ is a restriction operator on domain Ω, i.e.,

$$\begin{aligned}(P_+u)(x,y)&=u(x,y), &&\text{if } (x,y)\in\Omega\\ (P_+u)(x,y)&=0, &&\text{if } (x,y)\notin\overline{\Omega}.\end{aligned}$$

Let us change (7.3) by the perturbed equation

$$P_+K_\varepsilon p_+=f. \tag{7.4}$$

(In (7.3), (7.4) the constant C_1 is omitted; index "+" is retained for function p to ensure its existence in domain Ω).

This change is needed because the operator K is unbounded in H_s - space, although at the same time the operators K_ε have corresponding properties: they boundedly act from space $H_s(\mathbb{R}^m)$ into space $H_{s+1}(\mathbb{R}^m)$ according to Proposition 4.2.2.

The equation (7.4) will be solved explicitly for all $\varepsilon\neq 0$; if $p_+^{(\varepsilon)}$ is a solution of equation (7.4) and there exists $\lim\limits_{\varepsilon\to 0}p_+^{(\varepsilon)}=q$ in H_s - norm, then as one easily verifies, q will satisfy the equation (7.3). It allows us to call $p_+^{(\varepsilon)}$ an approximate solution of equation 7.3 (although a precise generally speaking does not exist a priori for an arbitrary right-hand side $f\in H_{s+1}(\Omega)$; here and below $H_s(\Omega)$ will denote the space of functions $u\in H_s(\mathbb{R}^2)$ for which $\operatorname{supp}u\subset\overline{\Omega}$).

Let O_1 be a cone in $\mathbb{R}^2$ of form $\{(\xi,\eta):\ a\eta>|\xi|\}$, O_2 be a reverse cone, $O_2=\{(\xi,\eta):\ (-\xi,-\eta)\in O_1\}$, $T(O_j)$, $j=1,2$, be radial tube domain over cone O_j.

Let us represent $\sigma(\xi,\eta,\varepsilon)$ in the form

$$\sigma(\xi,\eta,\varepsilon)=\sigma_1(\xi,\eta,\varepsilon)\sigma_2(\xi,\eta,\varepsilon), \tag{7.5}$$

where

$$\sigma_1(\xi,\eta,\varepsilon)=\left(\sqrt{a^2+1}\eta+\sqrt{a^2\eta^2-\xi^2-\varepsilon^2}\right)^{-1/2},$$

$$\sigma_2(\xi,\eta,\varepsilon)=\left(\sqrt{a^2+1}\eta-\sqrt{a^2\eta^2-\xi^2-\varepsilon^2}\right)^{-1/2},$$

and by $\sqrt{a^2\eta^2-\xi^2-\varepsilon^2}$ we mean a boundary value of the analytical in $T(O_1)$ function $\sqrt{a^2z_2^2-z_1^2-\varepsilon^2}$ of two compex variables z_1, z_2.

The representation (7.5) is a wave factorization of symbol $\sigma(\xi,\eta,\varepsilon)$ (see Chapter 5). Let us recall the properties of wave factorization elements which will be needed for furter discussion: $\sigma_1^{\pm 1}, \sigma_2^{\pm 2}$ admit analytical continuation into $T(O_1)$, $T(O_2)$ respectively, and these continuations satisfy the estimate from Definition 2.3.1 with $p=1$, $a=0$, $\alpha=0$, universal constant M and some $\beta \geq 0$ and have no zeroes in $T(O_1)$, $T(O_2)$ (for $\sigma_1^{-1}, \sigma_2^{-1}$ one can take $\beta = 1/2$, and for σ_1, σ_2 the $\beta = 0$); in other words $\sigma_1^{\pm 1}, \sigma_2^{\pm 1}$ belongs to classes $H_0(O_1)$, $H_0(O_2)$ respectively.

We will seek solution of equation (7.4) in the space $H_s(\Omega)$ of distributions u, and we will consider the right-hand side f of equation (7.4) in the space $H'_{s+1}(\Omega)$ of functions from $S'(\Omega)$ admitting a continuation into $H_{s+1}(\mathbb{R}^2)$, by definition

$$||f||_{H'_{s+1}}(\Omega) = \inf ||\ell f||_{s+1},$$

where the infimum is taken on all continuations ℓ.

Let f_1 be an arbitrary continuation of f from Ω on $\mathbb{R}^2$, $f_1 \in H_{s+1}(\mathbb{R}^2)$. Let

$$p_- = f_1 - K_\varepsilon p_+,$$

and rewrite the equation (7.4) in the form

$$K_\varepsilon p_+ + p_- = f_1 \qquad (7.6)$$

Applying to (7.6) the Fourier transform one can write

$$\sigma(\xi,\eta,\varepsilon)\tilde{p}_+(\xi,\eta) + \tilde{p}_-(\xi,\eta) = \tilde{f}_1(\xi,\eta).$$

Taking into account the wave factorization (7.5), the last equation can be expressed as

$$\begin{aligned}\sigma_1(\xi,\eta,\varepsilon)\tilde{p}_+(\xi,\eta) + \sigma_2^{-1}(\xi,\eta,\varepsilon)\tilde{p}_-(\xi,\eta) &= \\ = \sigma_2^{-1}(\xi,\eta,\varepsilon)\tilde{f}_1(\xi,\eta).\end{aligned} \qquad (7.7)$$

We introduce the notations:

$$\sigma_1(\xi,\eta,\varepsilon)\tilde{p}_+(\xi,\eta) = \tilde{q}_+(\xi,\eta),$$

$$\sigma_2^{-1}(\xi,\eta,\varepsilon)\tilde{p}_-(\xi,\eta) = \tilde{q}_-(\xi,\eta),$$

$$\sigma_2^{-1}(\xi,\eta,\varepsilon)\tilde{f}_1(\xi,\eta) = \tilde{f}_2(\xi,\eta),$$

and rewrite the equation (7.7):

$$\tilde{q}_+(\xi,\eta) + \tilde{q}_-(\xi,\eta) = \tilde{f}_2(\xi,\eta). \qquad (7.8)$$

(We "omit" the dependence $\tilde{q}_+$, $\tilde{q}_-$ on ε to simplify the notations).

The equation (7.8) is already well-known to us as a multidimensional jump problem (see Chapter 6 and Appendix 1), but in the more general context of functional spaces. Now we are convinced of its validity.

Let us denote by $\tilde{H}_s^+$ the Fourier-image of space $H_s(\Omega)$; $\tilde{H}_s^-$ is the Fourier-image of space $H_s(\mathbb{R}^2 \setminus \overline{\Omega})$, and $\tilde{H}_s$ is the Fourier-image of space $H_s(\mathbb{R}^2)$. Our first aim will be to verify that $\tilde{q}_\pm \in \tilde{H}^\pm_{s+1/2}$. Immediately one can note the following: $\tilde{q}_\pm \in \tilde{H}_{s+1/2}$, $\tilde{f}_2 \in \tilde{H}_{s+1/2}$ according to Proposition 4.1.1, because $p_+ \in H_s(\mathbb{R}^2)$ and the pseudodifferential operator with symbol σ_1 has order $-1/2$, the pseudodifferential operator with symbol σ_2^{-1} has order $1/2$ and $f_1, p_- \in H_{s+1}(\mathbb{R}^2)$. Further, the space $\tilde{H}^+_{s+1/2}$ has a precise description (Proposition 3.3.1): it consists of boundary values in the distribution sense of analytical in $T(O_1)$ functions $\tilde{u}(\xi,\eta)$ with finite norm

$$\sup\left(\iint\limits_{\mathbb{R}^2} |\tilde{u}(\xi+i\xi',\eta+i\eta')|^2 (1+|\xi|+|\eta|)^{2s+1} d\xi d\eta\right)^{1/2};$$

which coincides with quality

$$\left(\iint\limits_{\mathbb{R}^2} |\tilde{u}(\xi,\eta)|^2 (1+|\xi|+|\eta|)^{2s+1} d\xi d\eta\right)^{1/2}.$$

Sinse $\tilde{p}_+$ and σ_1 are boundary values of analytical in $T(O_1)$ functions, and

$$\iint\limits_{\mathbb{R}^2} |\sigma_1(\xi,\eta,\varepsilon)|^2 |\tilde{p}_+(\xi,\eta)|^2 (1+|\xi|+|\eta|)^{2s+1} d\xi d\eta \le$$

$$\le d\iint\limits_{\mathbb{R}^2} (1+|\xi|+|\eta|)^{-1} |\tilde{p}_+(\xi,\eta)|^2 (1+|\xi|+|\eta|)^{2s+1} d\xi d\eta =$$

$$= d\iint\limits_{\mathbb{R}^2} |\tilde{p}_+(\xi,\eta)|^2 (1+|\xi|+|\eta|)^{2s} d\xi d\eta,$$

however the last quantity is finite (because $\tilde{p}_+ \in \tilde{H}^+_s$), thus $\tilde{q}_+ \in \tilde{H}^+_{s+1/2}$.

Let us consider $\tilde{q}_-$. Since $\sigma_2^{-1} \in H_0(O_2)$, then by Proposition 2.3.2 in inverse Fourier transform in the distribution sense of function $\sigma_2^{-1}(\xi,\eta,\varepsilon)$ is the distribution b supported on $\overline{\Omega}_1\{(x,y) \in \mathbb{R}^2 : y \le -a|x|\}$, and $\sigma_2^{-1}\cdot\tilde{p}$ is a convolution of functions b and p_-. Approximating p_- by infinitely differentiable compact supports in $\mathbb{R}^2 \setminus \Omega$ functions and taking into account supp $(b * p_-) \subset \mathbb{R}^2 \setminus \Omega$, it is easily verified that supp $(b * p_-) \subset \mathbb{R}^2 \setminus \Omega$. The last is equivalent to $\sigma_2^{-1}\tilde{p}_- \in \tilde{H}^-_{s+1/2}$.

The equation (7.8) will be solved with the help of a specific integral operator which we now introduce (in a different form in comparison with Chapter 6) and we describe its basic properties.

On functions from the Schwartz class $S(\mathbb{R}^2)$, let us define the integral operator G_2' by the formula

$$(G_2'u)(\xi,\eta) = \lim_{\tau\to+0} \iint\limits_{\mathbb{R}^2} \frac{u(x,y)dxdy}{(\xi-x)^2 - a^2(\eta-y+i\tau)^2}.$$

If by $\Theta(x,y)$ we denote the characteristic function of set Ω, then it is not difficult to satisfy oneself that

$$F(\Theta \cdot u) = 2aG_2'\tilde{u}, \quad \forall u \in S(\mathbb{R}^2). \tag{7.9}$$

Since the operator of G_2' type in Appendix 1 is introduced with the help of Bochner kernel, we will carry out here and exact calculation.

Let us consider the integral ($\tau > 0$)

$$\iint\limits_{\mathbb{R}^2} e^{i(x\xi+y\eta)}\Theta(x,y)u(x,y)e^{-\tau y}\,dy;$$

it is a Fourier transform of the product of two functions $u(x,y)$ and $\Theta(x,y)e^{-\tau y}$ which are absolutely integrable (the last property allows us to apply a convolution theorem). Let us find a Fourier transform of function $\Theta(x,y)e^{-\tau y}$.

$$\iint\limits_{\mathbb{R}^2} e^{i(x\xi+y\eta)}\Theta(x,y)e^{-\tau y}\,dxdy = \iint\limits_{\Omega} e^{i(x\xi+y\eta)}e^{-\tau y}\,dxdy =$$

$$= \iint\limits_{\Omega} e^{ix\xi}e^{iy(\eta+i\tau)}\,dxdy = \int\limits_{-\infty}^{+\infty}\left(\int\limits_{a|x|}^{+\infty} e^{iy(\eta+i\tau)}\,dy\right)e^{ix\xi}\,dx =$$

$$= -\frac{1}{\eta+i\tau}\int\limits_{-\infty}^{+\infty} e^{ia|x|(\eta+i\tau)}e^{ix\xi}\,dx =$$

$$= -\frac{1}{i(\eta+i\tau)}\left(\int\limits_{-\infty}^{0} e^{-iax(\eta+i\tau)}e^{ix\xi}\,dx + \int\limits_{0}^{+\infty} e^{iax(\eta+i\tau)}e^{ix\xi}\,dx\right) =$$

$$= -\frac{1}{i(\eta+i\tau)}\left(\int\limits_{-\infty}^{0} e^{ix(\xi-a(\eta+i\tau))}\,dx + \int\limits_{0}^{+\infty} e^{ix(\xi+a(\eta+i\tau))}\,dx\right) =$$

$$= -\frac{1}{i(\eta+i\tau)}\left(\frac{1}{i(\xi-a(\eta+i\tau))} - \frac{1}{i(\xi+a(\eta+i\tau))}\right) =$$

$$= \frac{1}{\eta+i\tau}\frac{2a(\eta+i\tau)}{\xi^2-a^2(\eta+i\tau)^2} = \frac{2a}{\xi^2-a^2(\eta+i\tau)^2}.$$

The convolution of the last function with $\tilde{u}(\xi,\eta)$ and passage to the limit under $\tau \to +0$ give the formula (7.9).

The operator G_2' can be written in more usual form if one makes the change of variables

$$\begin{cases} \xi' = \xi - a\eta \\ \eta' = \xi + a\eta \end{cases} \qquad \begin{cases} x' = x - ay \\ y' = x + ay\,. \end{cases} \tag{7.10}$$

Then

$$(G_2'u)\left(\frac{\eta'+\xi'}{2},\frac{\eta'-\xi'}{2a}\right)=$$

$$=\frac{1}{2a}\lim_{\tau\to+0}\iint\limits_{\mathbb{R}^2}\frac{u\left(\frac{y'+x'}{2},\frac{y'-x'}{2a}\right)dx'dy'}{(\xi'-x'-i\tau)(\eta'-y'+i\tau)}.$$

The last expression is the product of two one-dimensional Cauchy type integrals and their limits can be easily evaluated. Let's denote by S_1, S_2, S_{12} the operators

$$(S_1u)(x',y')=\text{ v.p. }\frac{1}{\pi i}\int\limits_{-\infty}^{+\infty}\frac{u(\xi',y')d\xi'}{\xi'-x'},$$

$$(S_2u)(x',y')=\text{ v.p. }\frac{1}{\pi i}\int\limits_{-\infty}^{+\infty}\frac{u(x',\eta')d\eta'}{\eta'-y'},$$

$$S_{12}=S_1S_2.$$

We have [105]

$$\begin{aligned}(G_2'u)\left(\frac{\eta'+\xi'}{2},\frac{\eta'-\xi'}{2a}\right)&=\frac{\pi^2}{2a}\left(u_1(\xi',\eta')-\right.\\ \left.-(S_1u_1)(\xi',\eta')+(S_2u_1)(\xi',\eta')-(S_{12}u_1)(\xi',\eta')\right),&\end{aligned}\tag{7.11}$$

where $u_1(x,y)=u\left(\frac{y+x}{2},\frac{y-x}{2a}\right)$.

The representation (7.11) is convenient for our purposes. With its help the problem of boundedness for G_2' with respect to the norm in $\tilde{H}_s$ - space for $|s|<1/2$ easily reduces to the corresponding one-dimensional results because, for the change of variables (7.10), we have $1+|x'|+|y'|\sim 1+|x|+|y|$, $1+|\xi'|+|\eta'|\sim 1+|\xi|+|\eta|$, i.e., the fraction of quantities under consideration is bounded from above and below by positive constant. Indeed, boundedness of operator G_2' with respect to the norm of H_s - space is equivalent to boundedness the of operator

$$(G_2^0)(\xi,\eta)=\lim_{\tau\to+0}\iint\limits_{\mathbb{R}^2}\frac{u(x,y)(1+|x|+|y|))^s dxdy}{(1+|\xi|+|\eta|)^s((\xi-x)^2-a^2(\eta-y+i\tau)^2)}$$

in the space $L_2(\mathbb{R}^2)$.

Since according to (7.11) the operator G_2' is a linear combination of operators S_1, S_2 and its products (up to change of variables which leads to equivalent H_s - norm), then one needs to use Proposition 4.2.3.

Finally, the last important property of operator G_2'; it is an orthogonal projector of space $\tilde{H}_s$ on $\tilde{H}_s^+$ (it is appropriate to recall that $|s|<1/2$). If by Θ we denote the multiplication operator on function $\Theta(x,y)$, it is easy to verify that it is bounded with respect to norm $H_s(\mathbb{R}^2)$; then boundedness of G_2' with respect to norm $\tilde{H}_s$ follows from (7.9) and from the fact that $S(\mathbb{R}^2)$ is dense in $H_s(\mathbb{R}^2)$. Under $u(x,y)\in S(\mathbb{R}^2)$ we have $\operatorname{supp}\Theta(x,y)u(x,y)\subset\overline{\Omega}$, thus $\Theta\cdot u\in H_s(\Omega)$.

The closedness of subspace $H_s(\Omega)$ in $H_s(\mathbb{R}^2)$ follows from Proposition 3.2.1. So, G_2' is a bounded linear operator acting from $\tilde{H}_s$ into $\tilde{H}_s^+$. In the same way we are convinced that $I - G_2'$ acts boundedly from $\tilde{H}_s$ into $\tilde{H}_s^-$. A corollary of this property is the fact on the basis of which the equation (7.8) will be solved: namely, for $|s| < 1/2$ an arbitrary function $\tilde{u}(\xi, \eta) \in \tilde{H}_s$ is uniquely represented in the form

$$\tilde{u}(\xi, \eta) = \tilde{u}_1(\xi, \eta) + \tilde{u}_2(\xi, \eta),$$

where $\tilde{u}_1 \in \tilde{H}_s^+$, $\tilde{u}_2 \in \tilde{H}_s^-$, and

$$\tilde{u}_1 = G_2'\tilde{u}, \quad \tilde{u}_2 \overline{=} (I - G_2')\tilde{u}.$$

The equality

$$u = \Theta \cdot u + (1 - \Theta) \cdot u, \quad u \in S(\mathbb{R}^2) \tag{7.12}$$

is evident. Passing on to Fourier images according to (7.9) and extending (7.12) on continuity from $S(\mathbb{R}^2)$ on $H_s(\mathbb{R}^2)$, one can write

$$\tilde{u} = G_2'\tilde{u} + (I - G_2')\tilde{u}, \quad \tilde{u} \in \tilde{H}_s, \tag{7.13}$$

and denote

$$G_2'\tilde{u} = \tilde{u}_1, \quad (I - G_2')\tilde{u} = \tilde{u}_2.$$

Now we must satisfy ourselve that uniqueness of representation (7.13) holds. It is enough to show that if

$$\tilde{u}_1 + \tilde{u}_2 = 0, \tag{7.14}$$

then $\tilde{u}_1 \equiv 0$, $\tilde{u}_2 \equiv 0$.

Taking $g_1 \in C_0^\infty(\Omega)$, $g_2 \in C_0^\infty(\mathbb{R}^2 \setminus \overline{\Omega})$ (these classes are dense in H_s^+, H_s^- respectively) we have

$$\Theta g_1 = g_1, \quad \Theta g_2 = 0.$$

Approximating u_1 by functions of g_1 type, and u_2 by functions of g_2 type, and taking into account the properties of operator G_2' we obtain

$$G_2'\tilde{u}_1 = \tilde{u}_1, \quad G_2'\tilde{u}_2 = 0. \tag{7.15}$$

Analogously

$$(I - G_2')\tilde{u}_1 = 0, \quad (I - G_2')\tilde{u}_2 = \tilde{u}_2. \tag{7.16}$$

Applying to (7.14) the equalities (7.15), (7.16) we obtain the result desired.

Now according to the above the solution of equation (7.8) for $|s + 1/2| < 1/2$, i.e., $-1 < s < 0$, has the form

$$\tilde{q}_+(\xi, \eta) = \left(G_2'\tilde{f}_2\right)(\xi, \eta).$$

Hence, in Fourier images the solution of equation (7.4) in expanded form looks as follows:

$$\tilde{p}_+(\xi, \eta) = \sigma_1^{-1}(\xi, \eta, \varepsilon) \lim_{\tau \to +0} \iint\limits_{\mathbb{R}^2} \frac{\tilde{f}_1(x, y)dxdy}{\sigma_2(x, y, \varepsilon)((\xi - x)^2 - a^2(\eta - y + i\tau)^2)}, \tag{7.17}$$

where f_1 is an arbitrary continuation of f from Ω on $\mathbb{R}^2$, $f_1 \in H_{s+1}(\mathbb{R}^2)$, and p_+ does not depend on choice of this continuation (in this one can satisfy oneself by methods of Chapter 6).

Let us write (7.17) in symbolic form

$$\tilde{p}_+ = \sigma_1^{-1} G_2' \sigma_2^{-1} \tilde{f}_1. \tag{7.18}$$

Because of the boundedness of operator G_2' in $\tilde{H}_s$ for $|s| < 1/2$ and the boundedness of pseudodifferential operators of order 1/2 with symbols σ_1^{-1}, σ_2^{-1} acting from space $H_s(\mathbb{R}^2)$ into $H_{s-1/2}(\mathbb{R}^2)$, the a priori estimate of the solution holds

$$||p_+||_{H_s(\Omega)} \le d||\ell f||_{H_{s+1}(\mathbb{R}^2)} \le d'||f||_{H'_{s+1}(\Omega)};$$

the right-hand inequlity occurs by virtue of (7.18) and does not depend on the choice of continuation ℓf, thus ℓf can be chosen so that the inequality is satisfied.

Passage to the limit in formula (7.17) under $\varepsilon \to 0$ "is realized" very simply: it is enough to take in formula (7.17) $\varepsilon = 0$. The point is that all operators standing on the right-hand side of (7.17) are bounded in corresponding H_s - spaces, and therefore the formula (7.17) has meaning for $\varepsilon = 0$. This perturbation of the given equation is needed so that in all intermediate stages the appearing operators will be bounded.

Thus, we have proved

Theorem 7.1. *Let $-1 < s < 0$. Then the equation (7.3) for any right-hand side $f \in H'_{s+1}(\Omega)$ has unique solution $q \in H_s(\Omega)$ which in Fourier images can be written in the form*

$$\tilde{q}(\xi,\eta) = \sigma_1^{-1}(\xi,\eta,0) \lim_{\tau\to+0} \iint\limits_{\mathbb{R}^2} \frac{\tilde{f}_1(x,y)dxdy}{\sigma_2(x,y,0)((\xi-x)^2 - a^2(\eta - y + i\tau)^2)},$$

where f_1 is an arbitrary continuation of f from Ω on $\mathbb{R}^2$, $f_1 \in H_{s+1}(\mathbb{R}^2)$.

The a priopi estimate holds:

$$||q||_{H_s(\Omega)} \le c||f||_{H'_{s+1}(\Omega)}.$$

Remark 7.1. If Ω is a non-convex cone (angle is more than π) it is not difficult to observe that the method stated is applicable to this case without basic changes: we must introduce the operator G_2' with respect to cone $\mathbb{R}^2 \setminus \overline{\Omega}$, and wave factorization of σ has been constructed for cones related to $\mathbb{R}^2 \setminus \overline{\Omega}$.

Remark 7.2. The case $a = 0$ when Ω "degenerates" into a half-space is simpler, and it may be solved by an ordinary Wiener-Hopf method in which the factorization is realized on one variable, and the second variable plays the role of a parameter. $\sigma(\xi,\eta,\varepsilon)$ has been represented in the form

$$\sigma(\xi,\eta,\varepsilon) = \left(\eta + i\sqrt{\xi^2+\varepsilon^2}\right)^{-1/2} \left(\eta - i\sqrt{\xi^2+\varepsilon^2}\right)^{-1/2},$$

so that the factors admit analytical continuation on η under fixed ξ into upper and lower half-spaces, satisfy the estimates requested, and have no zeroes.

8. Equations in an infinite plane angle

8.1. MAIN RESULTS

In this section we will put these "semi-empirical" studies of Chapters 6,7 related to concrete applied problems, on a stable mathematical base and we will investigate the questions of solvability of a number of general classes of pseudodifferential equations in cones.

By a pseudodifferential equation we mean an equation of type

$$PAu_+ = f, \tag{8.1.1}$$

where $u_+ \in H_s(C^m)$, i.e., $u_+ \in H_s(\mathbb{R}^m)$, $u_+(x) = 0$, $x \notin C^m$; C^m is a convex cone in m - dimensional space $\mathbb{R}^m$, f is given on C^m, P is a restriction operator on C^m, A is an elliptic pseudodifferential operator with symbol $A(\xi)$ satisfying the inequality (5.1).

Definition 8.1.1. *By 0 - wave factorization of elliptic symbol $A(\xi) \in C_\alpha$ with respect to cone C^m we mean the wave factorization of this symbol in the sense of definition 5.1.*

Everewhere below in this section we will suggest that $m = 2$ (in this case one has only one convex cone on the plane, and it has the form C^a_+ up to rotation), and that elliptic symbol $A(\xi) \in C_\alpha$ admits a wave factorization with respect to cone C^a_+.

In general we will keep here the notation of Chapter 7 in which essentially the following two results were proved.

Lemma 8.1.1. *Let $|s| < 1/2$. Then operator G'_2 is bounded in the space $L_2(\mathbb{R}^2)$ with weight $(1+|x|)^s$, i.e., operator G^0_2 defined by formula*

$$(G^0_2)(x) = \lim_{\tau\to+0} \int\limits_{\mathbb{R}^2} \frac{f(y)(1+|y|)^s dy}{(1+|x|)^s((x_1-y_1)^2 - a^2(x_2-y_2+i\tau)^2)}$$

is bounded in the space $L_2(\mathbb{R}^2)$.

Let $\tilde{H}_s(\mathbb{R}^2)$, $\tilde{H}_s(C^a_+)$, $\tilde{H}_s(\mathbb{R}^2 \setminus \overline{C^a_+})$ denote the Fourier-images of spaces $H_s(\mathbb{R}^2)$, $H_s(C^a_+)$, $H_s(\mathbb{R}^2 \setminus \overline{C^a_+})$ respectively.

Lemma 8.1.2. *Let $|s| < 1/2$. The operator G'_2 is a bounded projector of space $\tilde{H}_s(\mathbb{R}^2)$ on the subspace $\tilde{H}_s(C^a_+)$. Any function $\tilde{f} \in \tilde{H}_s(\mathbb{R}^2)$ is uniquely represented in the form*

$$\tilde{f} = \tilde{f}_+ + \tilde{f}_-,$$

where $\tilde{f}_+ \in \tilde{H}_s(C^a_+)$, $\tilde{f}_- \in \tilde{H}_s(\mathbb{R}^2 \setminus \overline{C^a_+})$, and $\tilde{f}_+ = G'_2\tilde{f}$.

Lemma 8.1.3. *Let functions $B_{\neq}(\xi+i\tau)$, $B_{=}(\xi+i\tau)$ be analytical in $T(\overset{*}{C}{}^a_+)$ and $T(\overset{*}{C}{}^a_-)$ and satisfy estimates*

$$\begin{aligned} |B_{\neq}(\xi+i\tau)| &\le c_1(1+|\xi|+|\tau|)^\alpha, \quad \tau \in \overset{*}{C}{}^a_+, \\ |B_{=}(\xi+i\tau)| &\le c_2(1+|\xi|+|\tau|)^\alpha, \quad \tau \in \overset{*}{C}{}^a_-. \end{aligned} \tag{8.1.2}$$

Then the multiplication operator on $B_{\neq}(\xi)$ boundedly acts from space $\tilde{H}_s(C_+^a)$ into $\tilde{H}_{s-\alpha}(C_+^a)$, and the multiplication operator on $B_=(\xi)$ from space $\tilde{H}_s(\mathbb{R}^2\setminus\overline{C_+^a})$ into space $\tilde{H}_{s-\alpha}(\mathbb{R}^2\setminus\overline{C_+^a})$.

Proof. The fact that multiplication operators on functions $B_{\neq}(\xi)$, $B_=(\xi)$ boundedly act from spaces $\tilde{H}_s(C_+^a)$, $\tilde{H}_s(\mathbb{R}^2\setminus\overline{C_+^a})$ into space $\tilde{H}_{s-\alpha}(\mathbb{R}^2)$ is well-known due to proposition 4.1.1. Let's show that $B_{\neq}(\xi)\tilde{u}_+(\xi)\in\tilde{H}_{s-\alpha}(C_+^a)$ for any $\tilde{u}_+\in\tilde{H}_s(C_+^a)$.

The space $\tilde{H}_s(C_+^a)$ has explicit description (see Section 3.3), but we must keep in mind the spaces $\mathcal{L}_s^2$ and $\mathcal{H}_s$ must be replaced one with the other): $\tilde{u}_+\in\tilde{H}_s(C_+^a)$ if and only if $\tilde{u}_+(\xi+i\tau)$ is analytical in $T(\overset{*}{C}{}_+^a)$ and the quantity

$$\sup\int\limits_{\mathbb{R}^2}|\tilde{u}(\xi+i\tau)|^2(1+|\xi|)^{2s}d\xi,\quad \tau\in\overset{*}{C}{}_+^a,$$

is finite and coincides with

$$\int\limits_{\mathbb{R}^2}|\tilde{u}_+(\xi)|^2(1+|\xi|)^{2s}d\xi.$$

Then evidently, $B_{\neq}(\xi+i\tau)\tilde{u}_+(\xi+i\tau)$ is analytical in $T(\overset{*}{C}{}_+^a)$ and

$$\sup_{\tau\in\overset{*}{C}{}_+^b}\int\limits_{\mathbb{R}^2}|B_{\neq}(\xi+i\tau)\tilde{u}_+(\xi+i\tau)|^2\,(1+|\xi|)^{2(s-\alpha)}d\xi=$$

$$=\int\limits_{\mathbb{R}^2}|B_{\neq}(\xi)\tilde{u}_+(\xi)|^2\,(1+|\xi|)^{2(s-\alpha)}d\xi\le$$

$$\le c\int\limits_{\mathbb{R}^2}|\tilde{u}(\xi)|^2\,(1+|\xi|)^{2s}d\xi,$$

i.e., $B_{\neq}(\xi)\tilde{u}_+(\xi)\in\tilde{H}_{s-\alpha}(C_+^a)$.

Now let us consider $B_=(\xi)\tilde{u}_-(\xi)$. Let at first $u_-\in C_0^\infty(\mathbb{R}^2\setminus\overline{C_+^a})$. By virtue of estimate (8.1.2) $F^{-1}B_=\equiv b$ exists in the distribution sense and $\operatorname{supp} b\subset-\overline{C_+^a}$, as it follows from Proposition 2.3.2. Then $F^{-1}(B_=\tilde{u}_-)=b*u_-$. By definition 1.2.3 of convolution

$$(b*u_-)(x)=\left(b(y),\overline{u_-(x-y)}\right),$$

where $\overline{u_-(x-y)}$ is considered as a function on y (x is fixed), and notation $b(y)$ means that functional b acts on y – variable. Let us show that $(b*u_-)(x)=0$ under $x\in C_+^a$. Consider two cases: $y\in -C_+^a$ and $y\notin -C_+^a$. In the first case $x-y\in C_+^a$ and, thus $u_-(x-y)=0$ because $\operatorname{supp} u_-(x-y)\subset\mathbb{R}^2\setminus C_+^a$. In the second case $(b*u_-)$ vanishes because $y\notin\operatorname{supp} b$.

Transfer to general case $u_-\in H_s(\mathbb{R}^2\setminus\overline{C_+^a})$ is realized by virtue of density of class $C_0^\infty(\mathbb{R}^2\setminus\overline{C_+^a})$ in space $H_s(\mathbb{R}^2\setminus\overline{C_+^a})$.

So, it was shown that $B_=(\xi)\tilde{u}_-(\xi) \in \tilde{H}_{s-\alpha}(\mathbb{R}^2)$ and supp $F^{-1}(B_=\tilde{u}_-) \subset (\mathbb{R}^2 \setminus C_+^a)$. Hence, $B_=(\xi)\tilde{u}_-(\xi) \in \tilde{H}_{s-\alpha}(\mathbb{R}^2 \setminus \overline{C_+^a})$. ■

Let us denote $H_s'(C_+^a)$ the space of distributions $S'(C_+^a)$ admitting continuation ℓf on $\mathbb{R}^2$ which belongs to $H_s(\mathbb{R}^2)$. Norm in space $H_s'(C_+^a)$ is given as

$$||f||_s^+ = \inf_{\ell} ||\ell f||_s,$$

where infimum is taken among all continuations of f which belong to $H_s(\mathbb{R}^2)$.

Here and below the notation $A\tilde{u}$ denotes multiplication of symbol $A(\xi)$ on function $\tilde{u}(\xi)$, and Au denotes acting of pseudodifferential operator A on function u.

Theorem 8.1.1. *Let* $æ - s = \delta$, $|\delta| < 1/2$. *Then equation (8.1.1) for any right-hand side* $f \in H_{s-\alpha}'(C_+^a)$ *has unique solution* $u_+ \in H_s(C_+^a)$ *for which its Fourier transform is written in form*

$$\tilde{u}_+ = A_{\neq}^{-1} G_2' A_=^{-1} \tilde{\ell f},$$

where ℓf *is arbitrary continuation of* $f \in H_{s-\alpha}'(C_+^a)$ *on* $H_{s-\alpha}(\mathbb{R}^2)$.

A priori estimate holds

$$||u_+||_s \le c||f||_{s-\alpha}^+.$$

Proof. Let's denote

$$u_- = \ell f - Au_+. \tag{8.1.3}$$

Taking into account 0 - wave factorization after applying to (8.1.3) the Fourier transform we have

$$A_{\neq}(\xi)\tilde{u}_+(\xi) + A_=^{-1}(\xi)\tilde{u}_-(\xi) = A_=^{-1}(\xi)\tilde{\ell f}(\xi). \tag{8.14}$$

According to properties of 0 - wave factorization elements $A_{\neq}(\xi)$, $A_=(\xi)$ and on Proposition 4.1.1 and Lemma 8.1.3 we have $A_{\neq}(\xi)\tilde{u}_+(\xi) \in \tilde{H}_{s-æ}(C_+^a)$, $A_=^{-1}(\xi)\tilde{u}_-(\xi) \in \tilde{H}_{s-æ}(\mathbb{R}^2 \setminus \overline{C_+^a})$ (because $\tilde{u}_- \in \tilde{H}_{s-\alpha}(\mathbb{R}^2 \setminus \overline{C_+^a})$), $A_=^{-1}(\xi)\tilde{\ell f}(\xi) \in \tilde{H}_{s-æ}(\mathbb{R}^2)$, where æ is index of 0 - wave factorization. Since $|s - æ| < 1/2$ then on Lemma 8.1.2

$$A_{\neq}\tilde{u}_+ = G_2' A_=^{-1} \tilde{\ell f},$$

it implies

$$\tilde{u}_+ = A_{\neq}^{-1} G_2' A_=^{-1} \tilde{\ell f}.$$

A priori estimate:

$$||u_+||_s = ||\tilde{u}_+||_s \le c||G_2' A_=^{-1} \tilde{\ell f}||_{s-æ} \le$$

$$\le c||A_=^{-1} \tilde{\ell f}||_{s-æ} \le c||\tilde{\ell f}||_{s-æ} =$$

$$= c||\tilde{\ell f}||_{s-\alpha} \le c||f||_{s-\alpha}^+,$$

taking into account boundedness of operator G_2' in $\tilde{H}_s(\mathbb{R}^2)$ for $|s| < 1/2$ and boundedness of continuation operator ℓ. ■

Theorem 8.1.2. *Let $æ - s = n + \delta$, $n > 0$ be an integer, $|\delta| < 1/2$. Then the equation (8.1.1) for any right-hand side $f \in H'_{s-\alpha}(C^a_+)$ has solution $u_+ \in H_s(C^a_+)$, for which its Fourier transform is written by formula*

$$\begin{gathered}\tilde{u}_+(\xi) = A^{-1}_{\neq} Q G'_2 Q^{-1} A^{-1}_{=} \ell \tilde{f} + \\ + A^{-1}_{\neq}\left(\sum_{k=0}^{n-1}\left(\tilde{c}_k(\xi_1 - a\xi_2)^k(\xi_1 + a\xi_2)^k + \tilde{d}_k(\xi_1 + a\xi_2)(\xi_1 - a\xi_2)^k\right) + \right. \\ \left. + \sum_{k_1+k_2=0}^{n_\delta} a_{k_1k_2}(\xi_1 - a\xi_2)^{k_1}(\xi_1 + a\xi_2)^{k_2}\right),\end{gathered} \tag{8.1.5}$$

where c_k, d_k are arbitrary functions from $H_{s_k}(\mathbb{R}_-)$, $H_{s_k}(\mathbb{R}_+)$ respectively, $Q(\xi)$ is arbitrary elliptic polynomial of order n satisfying the estimate (5.1) with $\alpha = n$, $s_k = s - æ + k + 1/2$, $k = 0, 1, \ldots, n-1$, $a_{k_1k_2} \in \mathbb{C}$,

$$n_\delta = \begin{cases} n-1, & \delta > 0 \\ n-2, & \delta \le 0, \end{cases}$$

and the formula (8.1.5) describes all possible solutions of equation (8.1.1). A priori estimate holds

$$||u_+||_s \le c\left(||f||^+_{s-\alpha} + \sum_{k=0}^{n-1}\left([c_k]_{s_k} + [d_k]_{s_k}\right) + \sum_{k_1+k_2=0}^{n_\delta} |a_{k_1k_2}|\right).$$

Proof. By virtue of Proposition 4.1.1

$$Q^{-1}(\xi)A^{-1}_{=}(\xi)\ell\tilde{f}(\xi) \in \tilde{H}_{s-æ+n}(\mathbb{R}^2) = \tilde{H}_{-\delta}(\mathbb{R}^2).$$

In Lemma 8.1.2, $Q^{-1}(\xi)A^{-1}_{=}(\xi)\ell\tilde{f}(\xi)$ is uniquely represented in form

$$\begin{gathered}Q^{-1}(\xi)A^{-1}_{=}(\xi)\ell\tilde{f}(\xi) = G'_2 Q^{-1}(\xi)A^{-1}_{=}(\xi)\ell\tilde{f}(\xi) + \\ +(I - G'_2)Q^{-1}(\xi)A^{-1}_{=}(\xi)\ell\tilde{f}(\xi).\end{gathered}$$

Then

$$\begin{gathered}\tilde{g}(\xi) \equiv A^{-1}_{=}(\xi)\ell\tilde{f}(\xi) = Q(\xi)G'_2Q^{-1}(\xi)\tilde{g}(\xi) + \\ +Q(\xi)(I - G'_2)Q^{-1}(\xi)\tilde{g}(\xi).\end{gathered}$$

Let us rewrite the equality (8.1.4) in the form

$$\begin{gathered}A_{\neq}(\xi)\tilde{u}_+(\xi) - Q(\xi)G'_2Q^{-1}(\xi)\tilde{g}(\xi) = \\ = Q(\xi)(I - G'_2)Q^{-1}(\xi)\tilde{g}(\xi) - A^{-1}_{=}(\xi)\tilde{u}_-(\xi).\end{gathered} \tag{8.1.6}$$

The right-hand side of the equality belongs to $\tilde{H}_{-n-\delta}(\mathbb{R}^2 \setminus \overline{C^a_+})$, the left-hand ones to $\tilde{H}_{-n-\delta}(C^a_+)$. Taking inverse Fourier transform in (8.1.6) we obtain identity for two functions from classes $\tilde{H}_{-n-\delta}(C^a_+)$ and $\tilde{H}_{-n-\delta}(\mathbb{R}^2 \setminus \overline{C^a_+})$. Hence, each of them is a distribution from class $H_{-n-\delta}(\mathbb{R}^2)$ supported on ∂C^a_+. It is not difficult to determine its general form. Without loss of generality we will assume that

$\partial C_+^a = (\{0\} \times [0;+\infty)) \times ((-\infty;0) \times \{0\})$ (it reduces to this case by change of variables (7.10)). Very general form of distribution mentioned is such

$$\sum_{k=0}^{r_1} c_k(x_1)\delta^{(k)}(x_2) + \sum_{k=0}^{r_2} d_k(x_2)\delta^{(k)}(x_1) + \\ + \sum_{k_1+k_2=0}^{r_3} a_{k_1k_2}\delta^{(k_1)}(x_1)\delta^{(k_2)}(x_2), \tag{8.1.7}$$

according to proposition 1.3.3, where $c_k \in H_{s_k}(-\infty;0) \equiv H_{s_k}(\mathbb{R}_-)$, $d_k \in H_{s_k}(0;+\infty) \equiv H_{s_k}(\mathbb{R}_+)$, $\delta^{(k)}$ is derivative of k th order of Dirac δ - function.

The representation (8.1.7) is a corollary of the theorem on the general form of two-dimensional ditribution supported on a straight line. The number of sum members in (8.1.7) and parameters s_k must be such that every summand belongs to $H_{-n-\delta}(\mathbb{R}^2)$. For example,

$$||c_k\delta^{(k)}||^2_{-n-\delta} = \int\limits_{\mathbb{R}^2} |\tilde{c}_k(\xi_1)|^2 \, |\xi_2|^{2k}(1+|\xi|)^{-2n-2\delta}d\xi < +\infty.$$

In the Fubini theorem

$$||c_k\delta(k)||^2_s = \int\limits_{-\infty}^{+\infty} |\tilde{c}_k(\xi_1)|^2 \left(\int\limits_{-\infty}^{+\infty} |\xi_2|^{2k}(1+|\xi|)^{-2n-2\delta}d\xi_2 \right) d\xi_1,$$

and $H_{-n-\delta}$ - norm of $c_k\delta^{(k)}$ is finite iff the integral

$$\int\limits_{-\infty}^{+\infty} |\xi_2|^k(1+|\xi_1|+|\xi_2|)^{-2n-2\delta}d\xi$$

is finite, and it obviously converges only if $k = 0, 1, \ldots, n-1$. Hence,

$$||c_k\delta^{(k)}||^2_{-n-\delta} \sim \int\limits_{-\infty}^{+\infty} |\tilde{c}_k(\xi_1)|^2 \, (1+|\xi_1|)^{2k-2n-2\delta+1}d\xi_1,$$

$$c_k \in H_{s_k}(\mathbb{R}_-), \quad s_k = k-n-\delta+1/2 = s - æ + k + 1/2, \quad k = 0, 1, \ldots, n-1;$$

$$\left|\left| a_{k_1k_2}\delta^{(k_1)}\delta^{(k_2)} \right|\right|^2_{-n-\delta} = |a_{k_1k_2}|^2 \int\limits_{\mathbb{R}^2} |\xi_1|^{2k_1}|\xi_2|^{2k_2}(1+|\xi|)^{-2n-2\delta}d\xi \sim$$

$$\sim \int\limits_0^{+\infty} (1+t)^{-2n-2\delta+2k_1+2k_2+1}dt.$$

The last integral converges only if $2n + 2\delta - 2k_1 - 2k_2 - 1 > 1$, i.e. $k_1 + k_2 < n + \delta - 1$, and it implies $k_1 + k_2 = 0, 1, \ldots, n_\delta$, where

$$n_\delta = \begin{cases} n-1, & \delta > 0 \\ n-2, & \delta \leq 0. \end{cases}$$

Let us prove a priori estimates. We have

$$\left\| \frac{\sum\limits_{k_1+k_2=0}^{n_\delta} a_{k_1k_2}\xi_1^{k_1}\xi_2^{k_2}}{A_{\neq}(\xi)} \right\|_s^2 \leq$$

$$\leq \sum_{k_1+k_2=0}^{n_\delta} |a_{k_1k_2}|^2 \int\limits_{\mathbb{R}^2} (1+|\xi|)^{2s+2k_1+2k_2-2æ} d\xi \leq$$

$$\leq c \sum_{k_1+k_2=0}^{n_\delta} |a_{k_1k_2}|^2,$$

because $2s+2k_1+2k_2-2æ = 2(n-\delta)+2(k_1+k_2) < -2$, $k_1+k_2 = 0,1,\ldots,n_\delta$.

One obtains estimate for $A_{=}^{-1}QG_2'Q^{-1}A_{=}^{-1}\tilde{\ell f}$ one obtains in the same way as ones Theorem 8.1.1.

Finally,

$$\left\| \frac{\sum\limits_{k=0}^{n-1} \tilde{c}_k(\xi_1-a\xi_2)(\xi_1+a\xi_2)^k}{A_{\neq}(\xi)} \right\|_s^2 \leq$$

$$\leq \sum_{k=0}^{n-1} \int\limits_{\mathbb{R}^2} \frac{|\tilde{c}_k(\xi_1-a\xi_2)|^2 (\xi_1+a\xi_2)^{2k}(1+|\xi|)^{2s} d\xi}{|A_{\neq}(\xi)|^2} \leq$$

(we change $\xi_1 - a\xi_2 = \eta_1$, $\xi_1 + a\xi_2 = \eta_2$, $1+|\xi| \sim 1+|\eta|$)

$$\leq c \sum_{k=0}^{n-1} \int\limits_{\mathbb{R}^2} |\tilde{c}_k(\eta_1)|^2 |\eta_2|^{2k}(1+|\eta|)^{2s}(1+|\eta|)^{-2æ} d\eta \leq$$

$$\leq c \sum_{k=0}^{n-1} \int\limits_{-\infty}^{+\infty} |\tilde{c}_k(\eta_1)|^2 (1+|\eta_1|)^{2s_k} d\eta_1 = c \sum_{k=0}^{n-1} [c_k]_{s_k}^2,$$

and analogously for the $\sum\limits_{k=0}^{n-1} \tilde{d}_k(\xi_1+a\xi_2)(\xi_1-a\xi_2)^k$.

Taking into account Lemma 8.1.2 one can verify that any function of (8.1.5) type with arbitrary c_k, d_k, $a_{k_1k_2}$ is a solution of equation (8.1.1). ■

Let now $æ - s = n + \delta$, $n < 0$ is integer, $|\delta| < 1/2$. Since $s - \alpha > æ - \delta - \alpha$ then if $f \in H'_{s-\alpha}(C_+^a)$ we have $f \in H_{æ-s-\alpha}(C_+^a)$. By Theorem 8.1.1 there exists a unique solution $w_+ \in H_{æ-\delta}(C_+^a)$ of equation (8.1.1), and

$$\tilde{w}_+ = A_{\neq}^{-1}G_2'A_{=}^{-1}\tilde{\ell f}. \tag{8.1.8}$$

We put $A_{=}^{-1}\tilde{\ell f} = \tilde{g}$, take change of variables (7.10) and denote

$$(G'_2\tilde{g})\left(\frac{x_2+x_1}{2}, \frac{x_2-x_1}{2a}\right) = \tilde{h}(x_1,x_2),$$

$$\tilde{g}\left(\frac{y_2+y_1}{2}, \frac{y_2-y_1}{2a}\right) = \tilde{g}_1(y_1,y_2).$$

Then obviously

$$\tilde{h}(x_1,x_2) = \lim_{\tau\to+0} \int\limits_{\mathbb{R}^2} \frac{\tilde{g}_1(y_1,y_2)dy}{(x_1-y_1-ai\tau)(x_2-y_2+ai\tau)}.$$

Let us expand the kernel $(x_j - y_j + ib)^{-1}$ by formula ($b = \pm a\tau$)

$$\frac{1}{x_1-y_1+ib} = \frac{1}{(x_1-x_2-i+ib)\left(1-\frac{y_1-x_2-i}{x_1-x_2-i+ib}\right)} =$$

$$= \sum_{k=0}^{p-1} \frac{(y_1-x_2-i)^k}{(x_1-x_2-i+ib)^{k+1}} +$$

$$+ \frac{(y_1-x_2-i)^p}{(x_1-x_2-i+ib)^{p+1}\left(1-\frac{y_1-x_2-i}{x_1-x_2-i+ib}\right)} =$$

$$= \sum_{k=0}^{p-1} \frac{\Lambda^k(y_1,x_2)}{\Lambda_b^k(x_1,x_2)} + \frac{\Lambda^p(y_1,x_2)}{\Lambda_b^p(x_1,x_2)(x_1-y_1+ib)},$$

where $\Lambda_b(x_1,x_2) \equiv x_1 - x_2 - i + ib$, $\Lambda(x_1,x_2) = x_1 - x_2 - i$.

Analogously,

$$\frac{1}{x_2-y_2+ib} = \sum_{r=0}^{q} \frac{\underline{\Lambda}^r(y_2,x_1)}{\underline{\Lambda}_b^{r+1}(x_1,x_2)} +$$

$$+ \frac{\underline{\Lambda}^q(y_2,x_1)}{\underline{\Lambda}_b^q(x_2,x_1)(x_2-y_2+ib)}, \quad \underline{\Lambda}_b(x_2,x_1) = x_2 - x_1 + i + ib.$$

Then

$$\frac{1}{(x_1-y_1-ai\tau)(x_2-y_2+ai\tau)} =$$

$$= \sum_{k=0}^{p-1}\sum_{r=0}^{q-1} \frac{\Lambda^k(y_1,x_2)\underline{\Lambda}^r(y_2,x_1)}{\Lambda_{-a\tau}^{k+1}(x_1,x_2)\underline{\Lambda}_{a\tau}^{r+1}(x_2,x_1)} +$$

$$+ \sum_{k=0}^{p-1} \frac{\Lambda^k(y_1,x_2)\underline{\Lambda}^q(y_2,x_1)}{\Lambda_{-a\tau}^{k+1}(x_1,x_2)\underline{\Lambda}_{a\tau}^q(x_2,x_1)(x_2-y_2+ai\tau)} + \tag{8.1.9}$$

$$+ \sum_{r=0}^{q-1} \frac{\underline{\Lambda}^r(y_2,x_1)\Lambda^p(y_1,x_2)}{\Lambda_{-a\tau}^p(x_1,x_2)\underline{\Lambda}_{a\tau}^{r+1}(x_2,x_1)(x_1-y_1-ai\tau)} +$$

$$+\frac{\Lambda^p(y_1,x_2)\underline{\Lambda}^q(y_2,x_1)}{\Lambda^p_{-a\tau}(x_1,x_2)\underline{\Lambda}^q_{a\tau}(x_2,x_1)(x_1-y_1-ai\tau)(x_2-y_2+ai\tau)}.$$

Let $\tilde{g}_1(y_1,y_2)\in S(\mathbb{R}^2)$. Multiplying two parts of identity (8.1.9) on $\tilde{g}_1(y_1,y_2)$, integrating on y_1, y_2 and passing the to limit for $\tau\to+0$, we will obtain

$$\tilde{h}(x_1,x_2)=\sum_{k=0}^{p-1}\sum_{r=0}^{q-1}\Phi_{-k-1,\,-r-1}(x,x)\int\limits_{\mathbb{R}^2}\Phi_{k,r}(y,x)\tilde{g}_1(y)dy+$$

$$+\lim_{\tau\to+0}\sum_{k=0}^{p-1}\Phi_{-k-1,\,-q}(x,x)\int\limits_{\mathbb{R}^2}\frac{\Phi_{k,q}(y,x)\tilde{g}_1(y)dy}{x_2-y_2+ia\tau}+$$

$$+\lim_{\tau\to+0}\sum_{r=0}^{q-1}\Phi_{-p,\,-r-1}(x,x)\int\limits_{\mathbb{R}^2}\frac{\Phi_{p,r}(y,x)\tilde{g}_1(y)dy}{x_1-y_1-ia\tau}+$$

$$+\Phi_{-p,-q}(x,x)\lim_{\tau\to+0}\int\limits_{\mathbb{R}^2}\frac{\Phi_{p,q}(y,x)\tilde{g}_1(y)dy}{(x_1-y_1-ia\tau)(x_2-y_2+ia\tau)},$$

where $\Phi_{p,q}(y,x)=\Lambda^p(y_1,x_2)\underline{\Lambda}^q(y_2,x_1)=(y_1-x_2-i)^p(y_2-x_1+i)^q$ is polynomial of order no more than $p+q$ with respect to variables x_1, x_2, y_1, y_2.

Let us write

$$\begin{aligned}\tilde{h}=&\sum_{k=0}^{p-1}\sum_{r=0}^{q-1}\Phi_{-k-1,\,-r-1}\Pi'\Phi_{k,\,r}\tilde{g}'_1+\\&+\sum_{k=0}^{p-1}\Phi_{-k-1,\,-q}\Pi^{(2)}_{+}\Pi'_1\Phi_{k,\,q}\tilde{g}_1+\\&+\sum_{r=0}^{q-1}\Phi_{-p,\,-r-1}\Pi^{(1)}_{-}\Pi'_2\Phi_{p,\,r}\tilde{g}_1+\\&+\Phi_{-p,\,-q}G'_2\Phi_{p,\,q}\tilde{g}_1,\end{aligned}\tag{8.1.10}$$

where

$$\Pi^{(1)}_{\pm}\tilde{g}=\lim_{\tau\to+0}\int\limits_{-\infty}^{+\infty}\frac{\tilde{g}(\eta_1,\xi_2)d\eta_1}{\eta_1-\xi_1\pm i\tau},$$

$$\Pi^{(2)}_{\pm}\tilde{g}=\lim_{\tau\to+0}\int\limits_{-\infty}^{+\infty}\frac{\tilde{g}(\xi_1,\eta_2)d\eta_2}{\eta_2-\xi_2\pm i\tau},$$

$$\Pi'_j\tilde{g}=\int\limits_{-\infty}^{+\infty}\tilde{g}(\xi_1,\xi_2)d\xi_j,\quad j=1,2,$$

$$\Pi'\tilde{g}=\int\limits_{\mathbb{R}^2}\tilde{g}(\xi_1,\xi_2)d\xi.$$

The formula (8.1.10) was obtained with assumption that $\tilde{g}_1 \in S(\mathbb{R}^2)$. But it will be valid in such case when $\tilde{g}_1 \in \tilde{H}_{p+q-\delta}(\mathbb{R}^2)$. We will not dwell on this fact, only remarking that it is needed to apply Eskin's arguments [92,93] and to use density of $S(\mathbb{R}^2)$ in $\tilde{H}_{p+q-\delta}(\mathbb{R}^2)$.

Let us turn to formula (8.1.8). Making change of variables (7.10), putting $p+q=|n|$, introducing notation

$$\tilde{w}_+\left(\frac{x_2+x_1}{2}, \frac{x_2-x_1}{2a}\right) \equiv \tilde{W}_+(x_1,x_2),$$

$$A_{\neq}^{-1}\left(\frac{x_2+x_1}{2}, \frac{x_2-x_1}{2a}\right) \equiv a_{\neq}^{-1}(x_1,x_2),$$

applying to (8.1.8) the formula (8.1.10) (since $A_{=}^{-1}\ell\tilde{f} \in \tilde{H}_{|n|-\delta}(\mathbb{R}^2)$) we'll obtain

$$\tilde{W}_+(x) = \sum_{k=0}^{p-1}\sum_{r=0}^{q-1} \frac{\tilde{P}_{k,r}(x)}{a_{\neq}(x)\Phi_{k+1,\,r+1}(x,x)} +$$

$$+\sum_{k=0}^{p-1} \frac{\tilde{M}_{k,q}(x)}{\Phi_{k+1,\,q}(x,x)a_{\neq}(x)} +$$

$$+\sum_{r=0}^{q-1} \frac{\tilde{N}_{p,r}(x)}{\Phi_{p,\,r+1}(x,x)a_{\neq}(x)} + \tilde{U}_+(x),$$

where $\tilde{P}_{k,r}(x) = \Pi'\Phi_{k,r}\tilde{g}_1$, $\tilde{M}_{k,q}(x) = \Pi_+^{(2)}\Pi_1'\Phi_{k,q}\tilde{g}_1$, $\tilde{N}_{p,r}(x) = \Pi_-^{(1)}\Pi_2'\Phi_{p,r}\tilde{g}_1$, $\tilde{U}_+(x) = a_{\neq}^{-1}(x)\Phi_{-p,\,-q}(x,x)G_2'\Phi_{p,q}\tilde{g}_1$.

The function $\tilde{M}_{k,q}(x)$ has the form

$$\tilde{M}_{k,q}(x) = \lim_{\tau\to+0}\int_{\mathbb{R}^2} \frac{\Phi_{k,q}(y,x)\tilde{g}_1(y)dy}{x_1-y_1-ia\tau} =$$

$$= \sum_{|\sigma|+|\beta|=0}^{k+q} e_{\sigma,\beta}x^\sigma \lim_{\tau\to+0}\int_{\mathbb{R}^2} \frac{y^\beta \tilde{g}_1 dy}{x_1-y_1-ia\tau},$$

where σ, β are multi-indices.

Let us consider in more detail

$$\lim_{\tau\to+0}\int_{\mathbb{R}^2} \frac{y^\beta\tilde{g}_1(y)dy}{x_1-y_1-ia\tau} = \lim_{\tau\to+0}\int_{-\infty}^{+\infty} \frac{y_1^{\beta_1}\tilde{g}_{\beta_2}(y_1)dy_1}{x_1-y_1-ia\tau},$$

where $\tilde{g}_{\beta_2}(y_1) = \int\limits_{-\infty}^{+\infty} y_2^{\beta_2}\tilde{g}_1(y_1,y_2)dy_2$.

To the integral of Cauchy type one can give the following form:

$$\lim_{\tau\to+0}\int_{\mathbb{R}^2} \frac{y_1^{\beta_1}\tilde{g}_{\beta_2}(y_1)dy_1}{x_1-y_1-ia\tau} =$$

$$= \sum_{j=0}^{\ell-1} \frac{i}{(x_1 - i)^{j+1}} \int_{-\infty}^{+\infty} (y_1 - i)^j y_1^{\beta_1} \tilde{g}_{\beta_2}(y_1) dy_1 +$$

$$+ \lim_{\tau \to +0} \frac{1}{(x_1 - i)^\ell} \int_{-\infty}^{+\infty} \frac{(y_1 - i)^\ell y_1^{\beta_1} \tilde{g}_{\beta_2}(y_1) dy_1}{y_1 - x_1 - ia\tau},$$

and, hence

$$\lim_{\tau \to +0} \int_{\mathbb{R}^2} \frac{y^\beta \tilde{g}_1(y) dy}{x_1 - y_1 - ia\tau} = \sum_{j=0}^{\ell-1} \frac{i\lambda_j}{(x_1 - i)^{j+1}} + \tilde{g}_\beta(x_1), \qquad (8.1.11)$$

where

$$\tilde{g}_\beta(x_1) = \frac{1}{(x_1 - i)^\ell} \lim_{\tau \to +0} \int_{\mathbb{R}^2} \frac{y^\beta (y_1 - i)^\ell \tilde{g}_1(y) dy}{x_1 - y_1 - ia\tau},$$

$$\lambda_j = \int_{\mathbb{R}^2} (y_1 - i)^j y^\beta \tilde{g}_1(y) dy.$$

Put $\ell = |n| - |\beta| - 1$. (In this case for $|\beta| = |n| - 1$ in formula (8.1.11) $\sum$ vanishes). It is not difficult to satisfy onesesf that $\tilde{g}_\beta(x_1) \in \tilde{H}_{|n|-\delta-|\beta|-1/2}(\mathbb{R}_-)$ (see below a priori estimates). Thus,

$$\sum_{k=0}^{p-1} \frac{\tilde{M}_{k,q}(x)}{\Phi_{k+1,q}(x,x) a_{\neq}(x)} = \sum_{k=0}^{p-1} \sum_{|\sigma|+|\beta|=0}^{k+q} \frac{e_{\sigma,\beta} x^\sigma}{\Phi_{k+1,q}(x,x) a_{\neq}(x)} \sum_{j=0}^{\ell-1} \frac{i\lambda_j}{(x_1 - i)^{j+1}} +$$

$$+ \sum_{k=0}^{p-1} \sum_{|\sigma|+|\beta|=0}^{k+q} \frac{e_{\sigma,\beta} x^\sigma \tilde{g}_\beta(x_1)}{\Phi_{k+1,q}(x,x) a_{\neq}(x)}.$$

It is clear that for $\tilde{N}_{p,r}(x)$ an analogous representation is valid:

$$\sum_{r=0}^{q-1} \frac{\tilde{N}_{p,r}(x)}{\Phi_{p,r+1}(x,x) a_{\neq}(x)} = \sum_{r=0}^{q-1} \sum_{|\gamma|+|t|=0}^{r+p} \frac{b_{\gamma,t} x^\gamma}{\Phi_{p,r+1}(x,x) a_{\neq}(x)} \sum_{j=0}^{\ell'-1} \frac{i\mu_j}{(x_2 - i)^{j+1}} +$$

$$+ \sum_{r=0}^{q-1} \sum_{|\gamma|+|t|=0}^{r+p} \frac{b_{\gamma,t} x^\gamma \tilde{f}_t(x_2)}{\Phi_{p,r+1}(x,x) a_{\neq}(x)}.$$

Taking into acount proposition 4.1.1 and lemma 8.1.3 we conclude that

$$\tilde{U}_+(\xi_1 - a\xi_2, \xi_1 + a\xi_2) \in \tilde{H}_{|n|-\delta-æ}(C_+^a) = \tilde{H}_s(C_+^a).$$

Then one can write

$$\tilde{W}_+(x) = \tilde{U}_+(x) + \sum_{k=0}^{p-1} \sum_{r=0}^{q-1} \frac{\tilde{P}_{k,r}(x)}{a_{\neq}(x) \Phi_{k+1,r+1}(x,x)} +$$

$$+\sum_{k=0}^{p-1}\sum_{|\sigma|+|\beta|=0}^{k+q}\frac{e_{\sigma,\beta}x^{\sigma}}{\Phi_{k+1,q}(x,x)a_{\neq}(x)}\sum_{j=0}^{\ell}\frac{i\lambda_j}{(x_1-i)^{j+1}}+$$

$$+\sum_{r=0}^{q-1}\sum_{|\gamma|+|t|=0}^{r+p}\frac{b_{\gamma,t}x^{\gamma}}{\Phi_{p,r+1}(x,x)a_{\neq}(x)}\sum_{j=0}^{\ell'}\frac{i\mu_j}{(x_2-i)^{j+1}}+ \qquad (8.1.12)$$

$$+\sum_{k=0}^{p-1}\sum_{|\sigma|+|\beta|=0}^{k+q}\frac{e_{\sigma,\beta}x^{\sigma}\tilde{g}_{\beta}(x_1)}{\Phi_{k+1,q}(x,x)a_{\neq}(x)}+$$

$$+\sum_{r=0}^{q-1}\sum_{|\gamma|+|t|=0}^{r+p}\frac{b_{\gamma,t}x^{\gamma}\tilde{f}_t(x_2)}{\Phi_{p,r+1}(x,x)a_{\neq}(x)}.$$

By methods of book [93] (Theorem 7.3) it is not difficult to satisfy oneself that representation $\tilde{W}_+$ in form (8.1.12) is unique; from that follows solution of equation (8.1.1) with right-hand side $f \in H'_{s-\alpha}(C_+^a)$ belonging to $H_s(C_+^a)$ if and only if the following conditions are satisfied:

$$\tilde{P}_{k,r}(x)\equiv 0,\quad \tilde{g}_{\beta}(x_1)\equiv 0,\quad \tilde{f}_t(x_2)\equiv 0,\quad \lambda_j,\mu_j\equiv 0$$

for all admissible values k, r, β, t, j.

Coefficients of polynomial $\tilde{P}_{k,r}(x)$ (of order no more than $|n|-2$) have the form:

$$c'_{\sigma,\beta}=c_{\sigma,\beta}\int_{\mathbb{R}^2} y^{\beta}\tilde{g}_1(y)dy,$$

if

$$\Phi_{k,r}(x,y)=\sum_{|\sigma|+|\beta|=0}^{k+r}c_{\sigma,\beta}x^{\sigma}y^{\beta}.$$

By virtue of Fourier transform properties the condition $c'_{\sigma,\beta}=0$ is equivalent to

$$\left.\frac{\partial^{\beta_1}}{\partial x_1^{\beta_1}}\frac{\partial^{\beta_2}}{\partial x_2^{\beta_2}}g_1\right|_{x=0}=0,\quad |\beta|=0,1,\ldots,|n|-2.$$

The condition $\lambda_j=0$ is equivalent to

$$\left.\frac{\partial^{j+\beta_1}}{\partial x_1^{j+\beta_1}}\frac{\partial^{\beta_2}}{\partial x_2^{\beta_2}}g_1\right|_{x=0}=0,$$

and analogously with μ_j.

The conditions $\tilde{g}_{\beta}(x_1)=0$, $\tilde{f}_t(x_2)=0$ are equivalent to:

$$\lim_{\tau\to+0}\int_{\mathbb{R}^2}\frac{y^{\beta}(y_1-i)^{\ell}\tilde{g}_1(y)dy}{x_1-y_1-ia\tau}=0,$$

$$\lim_{\tau\to 0}\int\limits_{\mathbb{R}^2}\frac{y^t(y_2-i)^k\tilde{g}_1(y)dy}{x_2-y_2+ia\tau}=0,$$

and it leads to equivalent (taking into account the properties of Fourier transform and Cauchy type integral, see Sections 3.1, 4.2):

$$\left.\frac{\partial^{\beta_1+\ell}}{\partial x_1^{\beta_1+\ell}}\frac{\partial^{\beta_2}}{\partial x_2^{\beta_2}}g_1\right|_{\substack{x_1\le 0\\ x_2=0}}=0,$$

$$\left.\frac{\partial^{t_1}}{\partial x_1^{t_1}}\frac{\partial^{t_2+k}}{\partial x_2^{t_2+k}}g_1\right|_{\substack{x_2\ge 0\\ x_1=0}}=0.$$

Let us sum up the results obtained.

Theorem 8.1.3. *Let $æ - s = n+\delta$, $n\in\mathbb{Z}$, $n<0$, $|\delta|<1/2$. Then for any function $f\in H'_{s-\alpha}(C^a_+)$ a solution of equation (8.1.1) exists and its Fourier transform is represented in form (8.1.12) where one needs to put $x_1=\xi_1-a\xi_1$, $x_2=\xi_1+a\xi_2$.*

The functions $U_+(ay_1-y_2, ay_1+y_2)\in H_s(C^a_+)$, $g_\beta(ay_1-y_2)$, $f_\beta(ay_1-y_2)$ from class $H_{|n|-\delta-|\beta|-1/2}(B_\pm)$ respectively, where $B_+=\{(y_1,y_2):\ ay_1-y_2\le 0, ay_1+y_2=0\}$, $B_-=\{(y_1,y_2):\ ay_1-y_2=0,\ ay_1+y_2\ge 0\}$, $|\beta|=0,1,\ldots,n-1$, and the constants λ_j, μ_j, $j=0,1,\ldots,\ell$, which are contained in representation (8.1.12), are defined uniquely.

In order that the equation (8.1.1) has a solution from $H_s(C^a_+)$ it is necessary and sufficient to satisfy the conditions:

$$\left.\left(\frac{1}{a}\frac{\partial}{\partial y_1}-\frac{\partial}{\partial y_2}\right)^{\beta_1}\left(\frac{1}{a}\frac{\partial}{\partial y_1}+\frac{\partial}{\partial y_2}\right)^{\beta_2}A_=^{-1}\ell f(y)\right|_{y=0}=0,$$

$$\left.\left(\frac{1}{a}\frac{\partial}{\partial y_1}-\frac{\partial}{\partial y_2}\right)^{\beta_1}\left(\frac{1}{a}\frac{\partial}{\partial y_1}+\frac{\partial}{\partial y_2}\right)^{\beta_2}A_=^{-1}\ell f(y)\right|_{\substack{ay_1-y_2\le 0\\ ay_1+y_2=0}}=0,$$

$$\left.\left(\frac{1}{a}\frac{\partial}{\partial y_1}-\frac{\partial}{\partial y_2}\right)^{\beta_1}\left(\frac{1}{a}\frac{\partial}{\partial y_1}+\frac{\partial}{\partial y_2}\right)^{\beta_2}A_=^{-1}\ell f(y)\right|_{\substack{ay_1-y_2=0\\ ay_1+y_2\ge 0}}=0.$$

The following estimates hold:

$$||u_+||_s\le c||f||^+_{s-\alpha};$$

$$[g_\beta]_{n_\beta}\le c||f||^+_{s-\alpha},\quad [f_\beta]_{n_\beta}\le c||f||^+_{s-\alpha},$$
$$n_\beta=|n|-\delta-|\beta|-1/2,\quad |\beta|=0,1,\ldots,|n|-1;$$

$$|c'_{\sigma,\beta}|\le c||f||^+_{s-\alpha},$$

$$|\lambda_j|\le c||f||^+_{s-\alpha}\quad |\mu_j|\le c||f||^+_{s-\alpha},$$
$$|\beta|=0,1,\ldots,|n|-2,\quad j=0,1,\ldots,\ell.$$

Proof. It is left to prove a priori estimates only. Using proposition 4.1.1 we have

$$||u_+||_s = ||U_+||_s = ||\tilde{U}_+||_s = \left|\left|a_{\neq}^{-1}\Phi_{p,q}^{-1}G_2'\Phi_{p,q}A_{=}^{-1}\tilde{\ell f}\right|\right|_s \leq$$

$$\leq c\left|\left|\Phi_{p,q}^{-1}G_2'\Phi_{p,q}A_{=}^{-1}\tilde{\ell f}\right|\right|_{s-æ} \leq c\left|\left|G_2'\Phi_{p,q}A_{=}^{-1}\tilde{\ell f}\right|\right|_{s-æ+n} \leq$$

$$\leq c\left|\left|\Phi_{p,q}A_{=}^{-1}\tilde{\ell f}\right|\right|_{s-æ+n} \leq c\left|\left|A_{=}^{-1}\tilde{\ell f}\right|\right|_{s-æ} \leq c\left|\left|\tilde{\ell f}\right|\right|_{s-\alpha} =$$

$$= c\,||\ell f||_{s-\alpha} \leq c||f||^{+}_{s-\alpha}.$$

Further, if for example $0 < \delta < 1/2$ then

$$[g_\beta]_{n_\beta} = \left[(x_1 - i)^{-\ell}\Pi_{-}^{(1)}\Pi_2' y^\beta (y_1 - i)^\ell A_{=}^{-1}\tilde{\ell f}\right]_{n_\beta} \leq$$

$$\leq \left[\Pi_{-}^{(1)}\Pi_2' y^\beta (y_1 - i)^\ell A_{=}^{-1}\tilde{\ell f}\right]_{|n|-\delta-|\beta|-1/2-\ell} \leq$$

(since $0 < |n| - \delta - |\beta| - 1/2 - \ell = 1/2 - \delta < 1/2$)

$$\leq c\left[\Pi_2' y^\beta (y_1 - i)^\ell A_{=}^{-1}\tilde{\ell f}\right]_{1/2-\delta} \leq$$

(by Proposition 3.1.2)

$$\leq c\left[y^\beta (y_1 - i)^\ell A_{=}^{-1}\tilde{\ell f}\right]_{1-\delta} \leq \ldots \leq c||f||^{+}_{s-\alpha}.$$

Finally,

$$|c'_{\sigma,\beta}| \leq c\left|\int_{\mathbb{R}^2} y^\beta g_1(y)dy\right| \leq$$

$$\leq \int_{\mathbb{R}^2} (1 + |y|)^{|\beta|} \cdot |A_{=}^{-1}(y)| \cdot |\tilde{\ell f}(y)| dy \leq$$

$$\leq \int_{\mathbb{R}^2} (1 + |y|)^{|\beta|+æ-\alpha} |\tilde{\ell f}(y)| dy =$$

$$= c\int_{\mathbb{R}^2} (1 + |y|)^{|\beta|+æ-s}(1 + |y|)^{s-\alpha}|\tilde{\ell f}(y)| dy \leq$$

(the Cauchy unequality)

$$\leq c\left(\int_{\mathbb{R}^2} (1 + |y|)^{2(|\beta|+æ-s)} dy\right)^{1/2} ||\tilde{\ell f}||_{s-\alpha}.$$

The last intrgral is finite if and only if $2(|\beta|+æ-s) < -2$, i.e., $|\beta|-|n|+\delta < -1$. This implies the estimate needed.

Analogously the λ_j, μ_j are estimated. ■

Remark 8.1.1. It is not difficult to satisfy oneself that statements of theorem 8.1.1 – 8.1.3 in fact do not depend on the choice of continuation ℓf (see Chapter 7 and [92,93]).

Remark 8.1.2. The coefficients $a_{k_1k_2}$ and functions c_k, d_k in Theorem 8.1.2 in fact are not independent, but they are related by some "concordance conditions" (see also [132]). One can say the same about functions f_β, g_β and coefficients $c'_{\sigma,\beta}$ in the Theorem 8.1.3. These "concordance conditions" can be easily discovered, but we will not turn our attention here to this fact.

8.2. WEDGES OF CODIMENSION 2

Here we will generalize the results of Section 8.1 on the multidimensional situation when the singularity is a wedge of codimension 2. Since proofs of our statements will almost duplicate the calculations of previous Section 8.1 we will be bound by definitions and formulations of results only.

By a wedge of codimension 2 in m - dimensional space $\mathbb{R}^m$ we will mean the set of type

$$E_+^m = \{\xi = (\xi_1, \ldots, \xi_{m-1}, \xi_m) : \ \xi_m > a|\xi_{m-1}|, \ a > 0\}.$$

Definition 8.2.1. By $(m-2)$ - wave factorization of symbol $A(\xi) \in C_\alpha$ with respect to cone E_+^m mean its representation in form

$$A(\xi) = A_{\neq}(\xi)A_{=}(\xi),$$

such that the functions $A_{\neq}(\xi)$, $A_{=}(\xi)$ satisfy the conditions:

1) $A_{\neq}(\xi)$ is defined generally speaking on the set only

$$\{x \in \mathbb{R}^m : \ a^2x_m^2 \neq x_{m-1}^2\};$$

2) $A_{\neq}(\xi)$ admits an analytic continuation into $T(\overset{*}{K}{}_+^2)$ on variables ξ_{m-1}, ξ_m for almost all $\xi'' = (\xi_1, \ldots, \xi_{m-2})$, and this continuation satisfies the condition

$$\begin{aligned} &\left|A_{\neq}^{\pm 1}(\xi'', \xi_{m-1} + i\tau_{m-1}, \xi_m + i\tau_m)\right| \leq \\ &\leq c(1 + |\xi| + |\tau|)^{\pm æ}, \quad \forall \tau \in \overset{*}{K}{}_+^a, \end{aligned} \tag{8.2.1}$$

where $\overset{*}{K}{}_+^a = \{\tau = (\tau_{m-1}, \tau_m) \in \mathbb{R}^2 : \ a\tau_m > |\tau_{m-1}|\}$.

The function $A_{=}(\xi)$ must have analogous properties with exponents $\pm(\alpha - æ)$ instead of $\pm æ$ in conditions (8.2.1) and must admit an analytical continuation into radial tube domain $T(-\overset{*}{K}{}_+^2)$ over the cone $-\overset{*}{K}{}_+^2$.

Number æ we will call index of $(m-2)$ - wave factorization.

Example 8.2.1. (see also Chapter 5). Let $A(\xi) = \xi_1^2 + \cdots + \xi_m^2 + \varepsilon^2$, $\varepsilon \in \mathbb{R}\setminus\{0\}$, is symbol of operator $-\Delta + \varepsilon^2 I$. It is obvious $A(\xi) \in C_2^\infty$. The desired $(m-2)$ - wave factorization with respect to cone E_+^m looks such:

$$A_{\neq}(\xi) = \sqrt{a^2+1}\,\xi_m + \sqrt{a^2\xi_m^2 - \xi_{m-1}^2 - |\xi''|^2 - \varepsilon^2},$$

$$A_{=}(\xi)=\sqrt{a^2+1}\,\xi_m+\sqrt{a^2\xi_m^2-\xi_{m-1}^2-|\xi''|^2-\varepsilon^2},$$

where as before by $\sqrt{a^2\xi_m^2-\xi_{m-1}^2-|\xi''|^2-\varepsilon^2}$ one means the boundary value of analytical in $T(\overset{*}{K}{}_+^2)$ function

$$\sqrt{a^2z_m^2-z_{m-1}^2-|\xi''|^2-\varepsilon^2},$$

$$z_m=\xi_m+i\eta_m,\quad z_{m-1}=\xi_{m-1}+i\eta_{m-1},\quad T(\overset{*}{K}{}_+^2)\ni(\eta_{m-1},\eta_m)\to 0.$$

Index of $(m-2)$ - wave factorization in this case is equal to 1.

On class $S(\mathbb{R}^2)$ let's define the integral operator (see Appendix 1).

$$\left(G'_{m-2}u\right)(\xi)=\lim_{\tau\to+0}\int\limits_{\mathbb{R}^2}\frac{u(\xi'',\eta_{m-1},\eta_m)d\eta_{m-1}d\eta_m}{(\xi_m-\eta_{m-1})^2-a^2(\xi_m-\eta_m+i\tau)^2}.$$

Lemma 8.2.1. *Let $|s|<1/2$. Then operator G'_{m-2} is a bounded projector of space $\tilde{H}_s(\mathbb{R}^m)$ on subspace $\tilde{H}_s(E_+^m)$.*

Any function $\tilde{u}\in\tilde{H}_s(\mathbb{R}^m)$ is uniquely represented in form

$$\tilde{u}=\tilde{u}_++\tilde{u}_-.$$

where $\tilde{u}_+\in\tilde{H}_s(E_+^m)$, $\tilde{u}_-\in\tilde{H}_s(\mathbb{R}^m\setminus\overline{E_+^m})$, and

$$\tilde{u}_+=G'_{m-2}\tilde{u}.$$

As before in Section 8.1 let us introduce the space $H'_s(E_+^m)$ of distributions from $S'(E_+^m)$ admitting continuation ℓf on $\mathbb{R}^m$ which belongs to $H_s(\mathbb{R}^m)$. Norm in space $H'_s(E_+^m)$ we'll define by

$$||f||_s^+=\inf||\ell f||_s,$$

where the infimum is taken among continuations of f belonging to $H_s(\mathbb{R}^m)$.

Let us consider a pseudodifferential equation of (8.1.1) type in cone E_+^m

$$P_{m-2}Au_+=f, \tag{8.2.2}$$

where P_{m-2} is a restriction operator on E_+^m, A is an elliptic pseudodifferential operator with symbol $A(\xi)\in C_\alpha$ which admits $(m-2)$ - wave factorization with respect to wedge E_+^m with the index æ.

Theorem 8.2.1. *Let $|æ-s|<1/2$. Then equation (8.2.2) for arbitrary right-hand side $f\in H'_{s-\alpha}(E_+^m)$ has a unique solution $u_+\in H_s(E_+^m)$, for which its Fourier transform is written in form*

$$\tilde{u}(\xi)=A_{\neq}^{-1}(\xi)G'_{m-2}A_{=}^{-1}(\xi)\widetilde{\ell f}(\xi),$$

where ℓf is arbitrary continuation of f onto $H_{s-\alpha}(\mathbb{R}^m)$.

The a priori estimate holds

$$||u||_{+} \leq c||f||^{+}_{s-\alpha}.$$

As before we will denote H_s - norm of function u defined in $\mathbb{R}^{m-1}$ by

$$[u]_s^2 \equiv \int\limits_{\mathbb{R}^{m-1}} |\tilde{u}(\xi')|^2 (1+|\xi'|)^{2s} d\xi',$$

but of function v defined in $\mathbb{R}^{m-2}$ by

$$[[v]]_s^2 \equiv \int\limits_{\mathbb{R}^{m-2}} |\tilde{v}(\xi'')|^2 (1+|\xi''|)^{2s} d\xi''.$$

Theorem 8.2.2. *Let* $æ - s = n + \delta$, $n > 0$, $n \in \mathbb{Z}$, $|\delta| < 1/2$. *The equation (8.2.2) for arbitrary right-hand side* $f \in H'_{s-\alpha}(E^m_+)$ *has solutions* $u_+ \in H_s(E^m_+)$ *for which their Fourier transforms are written in form*

$$\tilde{u}_+(\xi) = A_{\neq}^{-1}(\xi) Q(\xi) G'_{m-2} Q^{-1}(\xi) A_{=}^{-1}(\xi) \widetilde{\ell f}(\xi) +$$

$$+A_{\neq}^{-1}(\xi)\left(\sum_{k=0}^{n-1}\left(\tilde{c}_k(\xi'', \xi_{m-1} - a\xi_m)(\xi_{m-1} + a\xi_m)^k + \right.\right.$$

$$\left. + \tilde{d}_k(\xi'', \xi_{m-1} + a\xi_m)(\xi_{m-1} - a\xi_m)^k\right) +$$

$$\left. + \sum_{k_1+k_2=0}^{n_\delta} a_{k_1k_2}(\xi'')(\xi_{m-1} - a\xi_m)^{k_1}(\xi_{m-1} + a\xi_m)^{k_2}\right),$$

where ℓf *is arbitrary continuation of* f *on* $H_{s-\alpha}(\mathbb{R}^m)$, $Q(\xi)$ *is arbitrary polynomial on* ξ_{m-1}, ξ_m *of order* n *from class* C_n^∞, c_k, d_k *are arbitrary functions from* $H_{s_k}(\mathbb{R}^{m-1}_-)$, $H_{s_k}(\mathbb{R}^{m-1}_+)$ *respectively,* $s_k = s - æ + k + 1/2$, $a_{k_1k_2}$ *are arbitrary functions from* $H_{s-æ+k_1+k_2+1}(\mathbb{R}^{m-2})$, $k = 0, 1, \ldots, n-1$,

$$n_\delta = \begin{cases} n-1, & \delta > 0 \\ n-2, & \delta \leq 0. \end{cases}$$

A priori estimate holds

$$||u_+||_s \leq c\left(||f||^+_{s-\alpha} + \sum_{k=0}^{n-1}\left([c_k]_{s_k} + [d_k]_{s_k}\right) + \right.$$

$$\left. + \sum_{k_1+k_2=0}^{n_\delta} [[a_{k_1k_2}]]_{s-æ+k_1+k_2+1}\right).$$

Theorem 8.2.3. *Let* $æ - s = n + \delta$, $n < 0$, $n \in \mathbb{Z}$, $|\delta| < 1/2$. *In order for equation (8.2.2) with right-hand side* $f \in H'_{s-\alpha}(E^m_+)$ *to have solution* $u_+ \in H_s(E^m_+)$, *it is necessary and sufficient to satisfy conditions*

$$\left(\frac{\partial}{\partial x_{m-1}} - \frac{1}{a}\frac{\partial}{\partial x_m}\right)^{\beta_1}\left(\frac{\partial}{\partial x_{m-1}} + \frac{1}{a}\frac{\partial}{\partial x_m}\right)^{\beta_2}(A^{-1}_{=}\ell f)(x)\Big|_{\substack{x_{m-1}=0\\ x_m=0}} = 0,$$

$$\left(\frac{\partial}{\partial x_{m-1}} - \frac{1}{a}\frac{\partial}{\partial x_m}\right)^{\beta_1}\left(\frac{\partial}{\partial x_{m-1}} + \frac{1}{a}\frac{\partial}{\partial x_m}\right)^{\beta_2}(A^{-1}_{=}\ell f)(x)\Big|_{\substack{ax_{m-1}-x_m\le 0\\ ax_{m-1}+x_m=0}} = 0,$$

$$\left(\frac{\partial}{\partial x_{m-1}} - \frac{1}{a}\frac{\partial}{\partial x_m}\right)^{\beta_1}\left(\frac{\partial}{\partial x_{m-1}} + \frac{1}{a}\frac{\partial}{\partial x_m}\right)^{\beta_2}(A^{-1}_{=}\ell f)(x)\Big|_{\substack{ax_{m-1}-x_m=0\\ ax_{m-1}+x_m\ge 0}} = 0,$$

where β *is multi-index,* $\beta = (\beta_1, \beta_2)$, $|\beta| = 0, 1, \ldots, |n| - 1$, $A^{-1}_{=}$ *is a pseudodifferential operator with symbol* $A^{-1}_{=}(\xi)$, *and the conditions do not depend on continuation* ℓf.

Remark 8.2.1. Let us note that in case $n < 0$ for the solution one can obtain a special representation of (8.1.12) type which implies the conditions formulated above, but this representation is very tedious, and we will not show it here.

9. General boundary value problems

9.1. THE DIRICHLET AND NEUMANN PROBLEMS

Here we will consider the question of correct statement of a boundary value problem for pseudodifferential equation in an angle on a plane which we will give on the basis of Theorem 8.1.2, and we will begin with Dirichlet and Neumann problems. Everywhere below in Section 9.1 we assume:

$$m = 2, \quad n = 1, \quad f = 0, \quad a = 1.$$

In this case according to Theorem 8.1.2 the general solution of equation (8.1.1) has the form

$$\tilde{u}_+(\xi_1, \xi_2) = A^{-1}_{\neq}(\xi_1, \xi_2)(\tilde{c}_0(\xi_1 - \xi_2) + \tilde{d}_0(\xi_1 + \xi_2)), \tag{9.1.1}$$

and our next step will be to give conditions on $\tilde{u}_+$ to define $\tilde{c}_0$, $\tilde{d}_0$ uniquely.

In formula (9.1.1) let us change variables

$$\begin{cases} \xi_1 - \xi_2 = x_1 \\ \xi_1 + \xi_2 = x_2, \end{cases}$$

and rewrite (9.1.1) in the form

$$\tilde{U}_+(x_1, x_2) = a_{\neq}^{-1}(x_1, x_2)((c(x_1) + d(x_2)), \qquad (9.1.2)$$

where the following notations are used:

$$\tilde{U}_+(x_1, x_2) = \tilde{u}_+\left(\frac{x_2 + x_1}{2}, \frac{x_2 - x_1}{2}\right),$$

$$a_{\neq}^{-1}(x_1, x_2) = A_{\neq}^{-1}\left(\frac{x_2 + x_1}{2}, \frac{x_2 - x_1}{2}\right),$$

$$\tilde{c}_0(x_1) \equiv c(x_1), \quad \tilde{d}_0(x_2) \equiv d(x_2).$$

Let us develop some formal calculations.

We will integrate (9.1.2) at first on x_1, and then on x_2 (assuming integrals needed exist). We will obtain the following system of linear integral equations for finding of functions $c(x_1)$, $d(x_2)$:

$$\begin{cases} \int\limits_{-\infty}^{+\infty} a_{\neq}^{-1}(x_1, x_2)c(x_1)dx_1 + b_1(x_2)d(x_2) = v_1'(x_2) \\ b_2(x_1)c(x_1) + \int\limits_{-\infty}^{+\infty} a_{\neq}^{-1}(x_1, x_2)d(x_2)dx_2 = v_2'(x_1), \end{cases}$$

$$b_1(x_2) = \int\limits_{-\infty}^{+\infty} a_{\neq}^{-1}(x_1, x_2)dx_1, \quad b_2(x_1) = \int\limits_{-\infty}^{+\infty} a_{\neq}^{-1}(x_1, x_2)dx_2.$$

Dividing two sides of the first equation by $b_1(x_2)$, and the second one by $b_2(x_1)$ (assuming b_1, b_2 everywhere are non-vanishing), we will obtain

$$\begin{cases} \int\limits_{-\infty}^{+\infty} K(x, y)c(x)dx + d(y) = v_1(y) \\ c(x) + \int\limits_{-\infty}^{+\infty} M(x, y)d(y)dy = v_2(x), \end{cases} \qquad (9.1.3)$$

where $K(x, y) = a_{\neq}^{-1}(x, y)b_1^{-1}(y)$, $M(x, y) = a_{\neq}^{-1}(x, y)b_2^{-1}(x)$.

Let us consider the first equation of system (9.1.3). Let us write

$$\int\limits_{-\infty}^{0} K(x, y)c(x)dx + \int\limits_{0}^{+\infty} K(x, y)c(x)dx + d(y) = v_1(y).$$

Making the change $x = -t$ in the first integral we obtain

$$\int\limits_{0}^{+\infty} K(-x, y)c(-x)dx + \int\limits_{0}^{+\infty} K(x, y)c(x)dx + d(y) = v_1(y).$$

Let $c_0(x)$ be a restriction $c(x)$ on $(0;+\infty)$, $c_1(x)$ is a restriction $c(-x)$ on $(0;+\infty)$ analogously we define $d_0(y)$, $d_1(y)$, $v_{10}(y)$, $v_{11}(y)$, $v_{20}(y)$, $v_{21}(y)$. In the same way we work with the second equation of system (9.1.3).

The system (9.1.3) then can be written in form

$$\left\{\begin{aligned} &\int_0^{+\infty} K_{11}(x,y)c_0(x)dx + \int_0^{+\infty} K_{12}(x,y)c_1(x)dx + d_0(y) = v_{10}(y)\\ &\int_0^{+\infty} K_{21}(x,y)c_1(x)dx + \int_0^{+\infty} K_{22}(x,y)c_0(x)dx + d_1(y) = v_{11}(y)\\ &\int_0^{+\infty} M_{11}(x,y)d_0(y)dy + \int_0^{+\infty} M_{12}(x,y)d_1(y)dy + c_0(x) = v_{20}(x)\\ &\int_0^{+\infty} M_{21}(x,y)d_1(y)dy + \int_0^{+\infty} M_{22}(x,y)d_0(y)dy + c_1(x) = v_{21}(x), \end{aligned}\right. \tag{9.1.4}$$

where the following notations are used: for all $x > 0$, $y > 0$ we put $K_{11}(x,y) = K(x,y)$, $K_{12}(x,y) = K(-x,y)$, $K_{21}(x,y) = K(-x,-y)$, $K_{22}(x,y) = K(x,-y)$; analogously the kernels $M_{ij}(x,y)$ are defined.

We will take one other hypothesis: we will assume that elliptic symbol $A(\xi_1,\xi_2)$ admits homogeneous wave factorization with the index æ. In this case the kernels $K(x,y)$, $M(x,y)$ (together with them the kernels $K_{ij}(x,y)$, $M_{ij}(x,y)$ have this property) are homogeneous of order -1 on variables x, y. It is not difficult to verify it. Indeed,

$$b_1(tx_2) = \int_{-\infty}^{+\infty} a_{\neq}^{-1}(x_1,tx_2)dx_1 = t\int_{-\infty}^{+\infty} a_{\neq}^{-1}(ty,tx_2)dy =$$

(change $x_1 = ty$)

$$= t\int_{-\infty}^{+\infty} t^{-æ}a_{\neq}^{-1}(y,x_2)dy = t^{1-æ}b_1(x_2),$$

because the function $a_{\neq}(x,y)$ is homogeneous of order æ.

Then

$$K(tx,ty) = a_{\neq}^{-1}(tx,ty)b_1^{-1}(ty) = t^{-æ}a_{\neq}^{-1}(x,y)t^{æ-1}b_1^{-1}(y) =$$

$$= t^{-1}a_{\neq}^{-1}(x,y)b_1^{-1}(y) = t^{-1}K(x,y),$$

and analogously with $M(x,y)$.

It is well-known that equations and systems of integral equations on a half- axis with homogeneous of order -1 kernels can be solved with the help of the Mellin transform [249] (see Appendix 3). Applying the Mellin transform (for function f

we will denote it by $\hat{f}$) to system (9.1.4) we obtain 4×4 - system of linear algebraic equations

$$\begin{cases} \hat{K}_{11}(\lambda)\hat{c}_0(\lambda)+\hat{K}_{12}(\lambda)\hat{c}_1(\lambda)+\hat{d}_0(\lambda)=\hat{v}_{10}(\lambda) \\ \hat{K}_{22}(\lambda)\hat{c}_0(\lambda)+\hat{K}_{21}(\lambda)\hat{c}_1(\lambda)+\hat{d}_1(\lambda)=\hat{v}_{11}(\lambda) \\ \hat{c}_0(\lambda)+\hat{M}_{11}(\lambda)\hat{d}_0(\lambda)+\hat{M}_{12}(\lambda)\hat{d}_1(\lambda)=\hat{v}_{20}(\lambda) \\ \hat{c}_1(\lambda)+\hat{M}_{22}(\lambda)\hat{d}_0(\lambda)+\hat{M}_{21}(\lambda)\hat{d}_1(\lambda)=\hat{v}_{21}(\lambda), \end{cases} \tag{9.1.5}$$

(where by $K_{ij}(x)$, $M_{ij}(x)$ one means $K_{ij}(1,x)$, $M_{ij}(x,1)$ respectively, assuming that $x^{-1/2}K_{ij}(x)$, $x^{-1/2}M_{ij}(x)\in L(0;+\infty)$) with respect to unknowns $\hat{c}_0(\lambda)$, $\hat{c}_1(\lambda)$, $\hat{d}_0(\lambda)$, $\hat{d}_1(\lambda)$.

The condition for unique solvability of system (9.1.5) for arbitrary right-hand side is in that the determinant

$$\Delta_A(\lambda)=\begin{vmatrix} \hat{K}_{11}(\lambda) & \hat{K}_{12}(\lambda) & 1 & 0 \\ \hat{K}_{22}(\lambda) & \hat{K}_{21}(\lambda) & 0 & 1 \\ 1 & 0 & \hat{M}_{11}(\lambda) & \hat{M}_{22}(\lambda) \\ 0 & 1 & \hat{M}_{12}(\lambda) & \hat{M}_{21}(\lambda) \end{vmatrix}$$

is different from zero under all permissible values of λ.

The condition $\Delta_A(\lambda)\neq 0$ further we will call Shapiro-Lopatinskii condition in analogy with the half-space case.

Now let us describe the problem which has just been solved. According to Fourier transform properties it looks as follows.

Let $\Gamma^+=\{x\in \mathbb{R}^2 : \ x_2>|x_1|\}$; finding the function $u_+(x)\in H_s(\Gamma^+)$ satisfying the equation

$$(Au_+)(x)=0, \quad x\in\Gamma^+, \tag{9.1.6}$$

and boundary conditions

$$u_+\Big|_{\substack{x_2=x_1\\ x_1>0}}=v_1, \quad u_+\Big|_{\substack{x_2=-x_1\\ x_1<0}}=v_2 \tag{9.1.7}$$

on angle sides $\partial\Gamma^+$, $\text{æ}-s=1+\delta$, $|\delta|<1/2$.

It is left "to watch" the functional spaces from which one must take the functions v_1, v_2. According to Proposition 3.1.2 and Theorem 8.1.2 the functions v_1, v_2 must be from classes $H_{s_0}(\mathbb{R}_-)$, $H_{s_0}(\mathbb{R}_+)$ respectively, where

$$s_0=s-\text{æ}+1/2=-1-\delta+1/2=-1/2-\delta.$$

So, (9.1.6), (9.1.7) is the Dirichlet problem for the pseudodifferential equation (9.1.6) when on the angle sides one gives the trace of solution u_+.

Before we pass to a statement of general boundary value problems in a plane sector we note that it is not difficult to write an algorithm for solution of the scheme above for the Neumann problem when on the angle sides one gives trace of normal derivative of solution u_+.

We will discuss briefly the problem when on one angle side one gives the Dirichlet condition, and on other one the Neumann condition.

The Dirichlet-Neumann problem: finding the function $u_+ \in H_s(\Gamma^+)$ satisfying the equation (9.1.6) and boundary conditions

$$u_+\Big|_{\substack{x_2 = x_1 \\ x_1 > 0}} = v_1, \quad \frac{\partial u_+}{\partial n}\Big|_{\substack{x_2 = -x_1 \\ x_1 < 0}} = v_2. \tag{9.1.8}$$

Since the general solution of equation (9.1.6) is given by the same formula (9.1.1) one needs only to define the functions $\tilde{c}_0$, $\tilde{d}_0$.

Obviously the second condition in (9.1.8) can be written in form

$$\left(\frac{\partial u_+}{\partial x_1} + \frac{\partial u_+}{\partial x_2}\right)\Big|_{\substack{x_2 = -x_1 \\ x_1 < 0}} = \sqrt{2}v_2. \tag{9.1.9}$$

Applying linear change of variables

$$\begin{cases} x_1' = x_2 + x_1 \\ x_2' = x_2 - x_1 \end{cases}$$

the condition (9.1.8) can be written in form

$$U_+\Big|_{\substack{x_2' = 0 \\ x_1' > 0}} = v_1', \quad \frac{\partial U_+}{\partial x_1'}\Big|_{\substack{x_1' = 0 \\ x_2' > 0}} = v_2'. \tag{9.1.10}$$

Now taking into account that the general solution we wrote in form (9.1.2) and using the fact that the Fourier transform commutes with rotations, we will apply it to (9.1.10). We obtain

$$\begin{aligned} &\int_{-\infty}^{+\infty} \tilde{U}_+(\xi_1, \xi_2)d\xi_2 = \tilde{v}_1'(\xi_1), \\ &\int_{-\infty}^{+\infty} \xi_1 \tilde{U}_+(\xi_1, \xi_2)d\xi_1 = \tilde{v}_2'(\xi_2) \end{aligned} \tag{9.1.10'}$$

Then for general solution (9.1.2) we will do the following. We integrate two parts (9.1.2) on x_2, then multiply two parts (9.1.2) on x_1 and integrate on x_1 (assuming as before that integrals needed are existing). We will obtain a system of (9.1.3) type, where the second equation will be the same, but in the first equation $K(x,y) = xa_{\neq}^{-1}(x,y)b_1^{-1}(y)$,

$$b_1(y) = \int_{-\infty}^{+\infty} xa_{\neq}^{-1}(x,y)dx.$$

Here it is not difficult to satisfy oneself also that in case of homogeneity of order æ for $a_{\neq}(x,y)$ the kernel $K(x,y)$ will be homogeneous of order -1 :

$$b_1(\lambda y) = \int_{-\infty}^{+\infty} xa_{\neq}^{-1}(x,\lambda y)dx = \int_{-\infty}^{+\infty} \lambda t a_{\neq}^{-1}(\lambda t x, \lambda y)\lambda dt =$$

$$= \lambda^{2-æ} \int\limits_{-\infty}^{+\infty} t a_{\neq}^{-1}(t,y)dt = \lambda^{2-æ} b_1(y);$$

$$K(\lambda x, \lambda y) = \lambda x a_{\neq}^{-1}(\lambda x, \lambda y) b_1^{-1}(\lambda y) =$$
$$= \lambda x \lambda^{-æ} \lambda^{æ-2} a_{\neq}^{-1}(x,y) x b_1^{-1}(y) = \lambda^{-1} K(x,y).$$

Further we work as in the previous case: we write out a system of two linear integral equations on the full axis $(-\infty; +\infty)$ as a 4×4 - system on the half-axis, apply a Mellin transform, compose determinant $\Delta_A(\lambda)$ and so on.

9.2. GENERAL BOUNDARY VALUE PROBLEMS

In this section we give a general statement of the boundary value problem starting from the form of solution which is prescribed by Theorem 8.1.2. Here in comparison with Section 9.1 we will conserve the restriction $m = 2$ only.

So, let $æ - s = n + \delta$, $n > 0$, $n \in \mathbb{Z}$, $|\delta| < 1/2$. Recall that by Theorem 8.1.2 a general solution of equation (8.1.1) has the form (8.1.5):

$$\begin{aligned} \tilde{u}_+(\xi) &= A_{\neq}^{-1} Q G_2' Q^{-1} A_{=}^{-1} \tilde{\ell f} + \\ &+ A_{\neq}^{-1} \left(\sum_{k=0}^{n-1} \tilde{c}_k(\xi_1 - a\xi_2)(\xi_1 + a\xi_2)^k + \right. \\ &\left. \tilde{d}_k(\xi_1 + a\xi_2)(\xi_1 - a\xi_2)^k \right). \end{aligned} \tag{9.2.1}$$

(Here we note at once that the form of solution (9.2.1) differ from (8.1.5), but is not a contradiction because the summands

$$\sum_{k_1+k_2=0}^{n_\delta} a_{k_1 k_2} (\xi_1 - a\xi_2)^{k_1} (\xi_1 + a\xi_2)^{k_2}$$

in general are contained in previous summands. Of course, one can work with the general form of solution (8.1.5), but then one needs to add the "concordance conditions" mentioned above).

The solution (9.2.1) depends on $2n$ arbitrary functions $c_k \in H_{s_k}(\mathbb{R}_-)$, $d_k \in H_{s_k}(\mathbb{R}_+)$, $s_k = s - æ + k + 1/2$, $k = 0, 1, \dots, n-1$, and to eliminate this non-uniqueness of solution we will give n boundary conditions on each angle side ∂C_+^a in following form:

$$(B_{1j} u_+)\Bigg|_{\substack{x_2 - a x_1 = 0 \\ x_1 > 0}} = g_{1j}, \quad (B_{2j} u_+)\Bigg|_{\substack{x_2 + a x_1 = 0 \\ x_1 < 0}} = g_{2j}, \tag{9.2.2}$$

where B_{rj}, $r = 1, 2$, are pseudodifferential operators with symbols $B_{rj}(\xi) \in C_{\beta_{rj}}^\infty$ which are homogeneous of order β_{rj} and satisfy the estimate

$$|B_{rj}(\xi)| \le c|\xi|^{\beta_{rj}}, \quad 0 < \beta_{rj} < s - 1/2. \tag{9.2.3}$$

Then by Proposition 4.1.1 $B_{rj}u_+$ belongs to $H_{s-\beta_{rj}}(\mathbb{R}^2)$ and by Proposition 3.1.2 the restriction B_{rj} on straight lines $x_2 \pm ax_1 = 0$ exists and belongs to $H_{s-\beta_{rj}-1/2}(\mathbb{R})$. Hence, the right-hand sides in (9.2.2) must be taken from classes $H_{s-\beta_{rj}-1/2}(\mathbb{R}_\pm)$.

Let us pass to Fourier images in (9.2.2) making simultaneously a linear change of variables which maps C_+^a onto the first quadrant. Then instead of (9.2.2) we have

$$\begin{aligned} \int\limits_{-\infty}^{+\infty} B'_{1j}(\xi)\tilde{U}_+(\xi)d\xi_2 = \tilde{g}'_{1j}(\xi_1), \\ \int\limits_{-\infty}^{+\infty} B'_{2j}(\xi)\tilde{U}_+(\xi)d\xi_1 = \tilde{g}'_{2j}(\xi_2), \end{aligned} \tag{9.2.4}$$

where

$$B'_{rj}(\xi) = B_{rj}\left(\frac{\xi_1-\xi_2}{2a}, \frac{\xi_1+\xi_2}{2}\right),$$

$$\tilde{U}_+(\xi) = \tilde{u}_+\left(\frac{\xi_1-\xi_2}{2a}, \frac{\xi_1+\xi_2}{2}\right).$$

The general solution, with the help of the same linear change of variables, we will write in form

$$\tilde{U}_+(\xi) = F(\xi) + a_{\neq}^{-1}(\xi)\sum_{k=0}^{n-1}\left(\tilde{c}_k(\xi_1)\xi_2^k + \tilde{d}_k(\xi_2)\xi_1^k\right),$$

$$F(\xi) = \left(A_{\neq}^{-1}QG'_2Q^{-1}A_{=}^{-1}\ell\tilde{f}\right)\left(\frac{\xi_1-\xi_2}{2a}, \frac{\xi_1+\xi_2}{2}\right),$$

and substitute in (9.2.4). Then we obtain the following

$$\begin{aligned} F_{1j}(\xi_1) + \sum_{k=0}^{n-1}\left[b_{1jk}(\xi_1)\tilde{c}_k(\xi_1) + \xi_1^k\int\limits_{-\infty}^{+\infty} K'_{1j}(\xi_1,\xi_2)\tilde{d}_k(\xi_2)d\xi_2\right] = \\ = \tilde{g}'_{1j}(\xi_1), \end{aligned} \tag{9.2.5}$$

where

$$F_{1j}(\xi_1) = \int\limits_{-\infty}^{+\infty} B'_{1j}(\xi)F(\xi)d\xi_2,$$

$$b_{1jk}(\xi_1) = \int\limits_{-\infty}^{+\infty} B'_{1j}(\xi)a_{\neq}^{-1}(\xi)\xi_2^k d\xi_2,$$

$$K'_{1j}(\xi) = B'_{1j}(\xi)a_{\neq}^{-1}(\xi)$$

on one angle side, and analogously on other one.

Let us write (9.2.5) in form

$$\begin{aligned}&\sum_{k=0}^{n-1} b_{1jk}(\xi_1)\left[\tilde{c}_k(\xi_1) + \int_{-\infty}^{+\infty} K_{1jk}(\xi_1,\xi_2)\tilde{d}_k(\xi_2)d\xi_2\right] = \\ &= \tilde{g}'_{1j}(\xi_1) - F_{1j}(\xi_1), \\ &K_{1jk}(\xi_1,\xi_2) = \xi_1^k b_{1jk}^{-1}(\xi_1,\xi_2)K'_{1j}(\xi_1,\xi_2).\end{aligned} \tag{9.2.6}$$

Let us write out an analogue of equality (9.2.6) for the other angle side:

$$\begin{aligned}F_{2j}(\xi_2) - \sum_{k=0}^{n-1}\Bigg[\xi_2^k \int_{-\infty}^{+\infty} K'_{2j}(\xi_1,\xi_2)\tilde{c}_k(\xi_1)d\xi_1 + \\ + b_{2jk}(\xi_2)\tilde{d}_k(\xi_2)\Bigg] = \tilde{g}'_{2j}(\xi_2),\end{aligned} \tag{9.2.5'}$$

and, hence,

$$\begin{aligned}&\sum_{k=0}^{n-1} b_{2jk}(\xi_2)\left[\int_{-\infty}^{+\infty} K'_{2jk}(\xi_1,\xi_2)\tilde{c}_k(\xi_1)d\xi_1 + \tilde{d}_k(\xi_2)\right] = \\ &= \tilde{g}'_{2j}(\xi_2) - F_{2j}(\xi_2).\end{aligned} \tag{9.2.6'}$$

In general the system of $2n$ linear integral equations with $2n$ unknowns $\tilde{c}_k(\xi_1)$, $\tilde{d}_k(\xi_2)$, $k = 0,1,\ldots,n-1$, one solves with great difficulty, and we will consider the special case when the right-hand sides of this system are subordinated to some restictions.

At first, let us verify that kernels K_{rjk} are homogeneous of order -1. In fact ($r = 1$),

$$b_{1jk}(\lambda\xi_1) = \int_{-\infty}^{+\infty} B'_{1j}(\lambda\xi_1,\xi_2)a_{\neq}^{-1}(\lambda\xi_1,\xi_2)\xi_2^k d\xi_2 =$$

$$= \int_{-\infty}^{+\infty} B'_{1j}(\lambda\xi_1,\lambda\xi_2)a_{\neq}^{-1}(\lambda\xi_1,\lambda\xi_2)(\lambda\xi_2)^k\lambda d\xi_2 =$$

$$= \lambda^{\beta_{1j}}\lambda^{-æ}\lambda^k\lambda b_{1jk}(\xi_1) = \lambda^{\beta_{1j}+æ-k-1}b_{1jk}(\xi_1);$$

$$K_{1jk}(\lambda\xi_1,\lambda\xi_2) = (\lambda\xi_1)^k\lambda^{-\beta_{1j}+æ-k-1}b_{1jk}(\xi)B'_{1j}(\lambda\xi)a_{\neq}^{-1}(\lambda\xi) =$$

$$= \lambda^{-\beta_{1j}+æ-k-1}\lambda^{\beta_{1j}}\lambda^{-æ}\xi_1^k b_{1jk}(\xi)B'_{1j}(\xi)a_{\neq}^{-1}(\xi) = \lambda^{-1}K_{1jk}(\xi).$$

Let us assume that the following conditions are satisfied:

$$\det\left(b_{1jk}(\xi_1)\right) \neq 0, \tag{9.2.7}$$

$$\det\left(b_{2jk}(\xi_2)\right) \neq 0, \tag{9.2.8}$$

for all $\xi_1, \xi_2 \neq 0$, and for $k = 0, 1, \ldots, n-1$,

$$\tilde{c}_k(\xi_1) + \int_{-\infty}^{+\infty} K_{1jk}(\xi_1, \xi_2)\tilde{d}_k(\xi_2)d\xi_2 = \tilde{c}_k(\xi_1) + \int_{-\infty}^{+\infty} K_{1kk}(\xi_1, \xi_2)\tilde{d}_k(\xi_2)d\xi_2,$$
$$j = 0, 1, \ldots, n-1,$$

i.e.,

$$\int_{-\infty}^{+\infty} K_{1jk}(\xi_1, \xi_2)\tilde{d}_k(\xi_2)d\xi_2 = \int_{-\infty}^{+\infty} K_{1kk}(\xi_1, \xi_2)\tilde{d}_k(\xi_2)d\xi_2, \tag{9.2.9}$$

and, analogously,

$$\int_{-\infty}^{+\infty} K_{2jk}(\xi_1, \xi_2)\tilde{c}_k(\xi_1)d\xi_1 = \int_{-\infty}^{+\infty} K_{2kk}(\xi_1, \xi_2)\tilde{c}_k(\xi_1)d\xi_1, \tag{9.2.10}$$
$$j = 0, 1, \ldots, n-1.$$

We rewrite now the system (9.2.6), (9.2.6′) in form

$$\sum_{k=0}^{n-1} b_{1jk}(\xi_1)a_{1k}(\xi_1) = \tilde{g}'_{1j}(\xi_1) - F_{1j}(\xi_1), \tag{9.2.11}$$

$$\sum_{k=0}^{n-1} b_{2jk}(\xi_2)a_{2k}(\xi_2) = \tilde{g}'_{2j}(\xi_2) - F_{2j}(\xi_2), \tag{9.2.11′}$$

where

$$a_{1k}(\xi_1) = \tilde{c}_k(\xi_1) + \int_{-\infty}^{+\infty} K_{1kk}(\xi_1, \xi_2)\tilde{d}_k(\xi_2)d\xi_2,$$

$$a_{2k}(\xi_2) = \tilde{d}_k(\xi_2) + \int_{-\infty}^{+\infty} K_{2kk}(\xi_1, \xi_2)\tilde{c}_k(\xi_1)d\xi_1.$$

By satisfying conditions (9.2.7), (9.2.8) (which are namely Shapiro-Lopatinskii conditions on angle sides; *cf.* Proposition 4.4.4) the system (9.2.11), (9.2.11′) is uniquely solvable with respect to a_{1k}, a_{2k} with arbitrary right-hand sides.

Let us denote $b^{-1}_{1kj}(\xi_1)$, $b^{-1}_{2kj}(\xi_2)$ the elements of matrices which are inverse for matrices $(b_{1kj}(\xi_1))$, $(b_{2kj}(\xi_2))$ respectively. We have

$$\begin{cases} \tilde{c}_k(\xi_1) + \int_{-\infty}^{+\infty} K_{1kk}(\xi_1, \xi_2)\tilde{d}_k(\xi_2)d\xi_2 = h_{1k}(\xi_1) \\ \int_{-\infty}^{+\infty} K_{2kk}(\xi_1, \xi_2)\tilde{c}_k(\xi_1)d\xi_1 + \tilde{d}_k(\xi_2) = h_{2k}(\xi_2) \end{cases} \tag{9.2.12}$$

where

$$h_{rk}(x) = \sum_{j=0}^{n-1} b_{rkj}^{-1}(x) \left(\tilde{g}'_{rj}(x) - F_{rj}(x)\right),$$
$$r = 1, 2, \quad k = 0, 1, \ldots, n-1.$$

Under every fixed k the system (9.2.12) is exactly the system (9.1.3), and for it according to Section 9.1 we introduce the functions $c_k^{(0)}$, $c_k^{(1)}$, $d_k^{(0)}$, $d_k^{(1)}$, $h_{1k}^{(0)}$, $h_{1k}^{(1)}$, $h_{2k}^{(0)}$, $h_{2k}^{(1)}$ and the kernels $K_{rkk}^{(ij)}$, $i, j = 1, 2$, write it as 4×4 – system and apply a Mellin transform. The matrix of linear algebraic system of equations will have the form

$$\hat{A}_k(\lambda) = \begin{pmatrix} \hat{K}_{2kk}^{(11)}(\lambda) & \hat{K}_{2kk}^{(12)}(\lambda) & 1 & 0 \\ \hat{K}_{2kk}^{(22)}(\lambda) & \hat{K}_{2kk}^{(21)}(\lambda) & 0 & 1 \\ 1 & 0 & \hat{K}_{1kk}^{(11)}(\lambda) & \hat{K}_{1kk}^{(22)}(\lambda) \\ 0 & 1 & \hat{K}_{1kk}^{(12)}(\lambda) & \hat{K}_{1kk}^{(21)}(\lambda) \end{pmatrix}.$$

If $\det \hat{A}_k(\lambda) \neq 0$, $\forall\lambda$, $\operatorname{Re}\lambda = 1/2$, $k = 0, 1, \ldots, n-1$, then

$$\hat{C}_k(\lambda) = \hat{A}_k^{-1}(\lambda)\hat{H}_k(\lambda), \tag{9.2.13}$$

where $\hat{C}_k(\lambda)$ is a vector with components $\hat{c}_k^{(0)}(\lambda)$, $\hat{c}_k^{(1)}(\lambda)$, $\hat{d}_k^{(0)}(\lambda)$, $\hat{d}_k^{(1)}(\lambda)$, $\hat{H}_k(\lambda)$ is a vector with components $\hat{h}_{2k}^{(0)}(\lambda)$, $\hat{h}_{2k}^{(1)}(\lambda)$, $\hat{h}_{1k}^{(0)}(\lambda)$, $\hat{h}_{1k}^{(1)}(\lambda)$.

Let us write the condition (9.2.9) in terms of the Mellin transform. For this we represent (9.2.9) as

$$\int_0^{+\infty} K_{1jk}^{(11)}(\xi_1, \xi_2) d_k^{(0)}(\xi_2) d\xi_2 + \int_0^{+\infty} K_{1jk}^{(12)}(\xi_1, \xi_2) d_k^{(1)}(\xi_2) d\xi_2 =$$

$$= \int_0^{+\infty} K_{1kk}^{(11)}(\xi_1, \xi_2) d_k^{(0)}(\xi_2) d\xi_2 + \int_0^{+\infty} K_{1kk}^{(12)}(\xi_1, \xi_2) d_k^{(1)}(\xi_2) d\xi_2;$$

$$\int_0^{+\infty} K_{1jk}^{(21)}(\xi_1, \xi_2) d_k^{(0)}(\xi_2) d\xi_2 + \int_0^{+\infty} K_{1jk}^{(22)}(\xi_1, \xi_2) d_k^{(1)}(\xi_2) d\xi_2 =$$

$$= \int_0^{+\infty} K_{1kk}^{(21)}(\xi_1, \xi_2) d_k^{(0)}(\xi_2) d\xi_2 + \int_0^{+\infty} K_{1kk}^{(22)}(\xi_1, \xi_2) d_k^{(1)}(\xi_2) d\xi_2,$$

and pass to Mellin images:

$$\hat{K}_{1jk}^{(11)}(\lambda)\hat{d}_k^{(0)}(\lambda) + \hat{K}_{1jk}^{(12)}\hat{d}_k^{(1)}(\lambda) =$$

$$= \hat{K}_{1kk}^{(11)}(\lambda)\hat{d}_k^{(0)}(\lambda) + \hat{K}_{1kk}^{(12)}\hat{d}_k^{(1)}(\lambda);$$

$$\hat{K}_{1jk}^{(21)}(\lambda)\hat{d}_k^{(0)}(\lambda) + \hat{K}_{1jk}^{(22)}(\lambda)\hat{d}_k^{(1)}(\lambda) =$$

$$= \hat{K}^{(21)}_{1kk}(\lambda)\hat{d}^{(0)}_k(\lambda) + \hat{K}^{(22)}_{1kk}(\lambda)\hat{d}^{(1)}_k(\lambda).$$

We write the last two equalities in form

$$\left(\hat{K}^{(11)}_{1jk}(\lambda) - \hat{K}^{(11)}_{1kk}(\lambda)\right)\hat{d}^{(0)}_k(\lambda) + \left(\hat{K}^{(12)}_{1jk}(\lambda) - \hat{K}^{(12)}_{1kk}(\lambda)\right)\hat{d}^{(1)}_k(\lambda) = 0,$$

and analogously for $\hat{c}^{(0)}_k(\lambda)$, $\hat{c}^{(1)}_k(\lambda)$:

$$\left(\hat{K}^{(11)}_{2jk}(\lambda) - \hat{K}^{(11)}_{2kk}(\lambda)\right)\hat{c}^{(0)}_k(\lambda) + \left(\hat{K}^{(12)}_{2jk}(\lambda) - \hat{K}^{(12)}_{2kk}(\lambda)\right)\hat{c}^{(1)}_k(\lambda) = 0;$$

$$\left(\hat{K}^{(21)}_{1jk}(\lambda) - \hat{K}^{(21)}_{1kk}(\lambda)\right)\hat{d}^{(0)}_k(\lambda) + \left(\hat{K}^{(22)}_{1jk}(\lambda) - \hat{K}^{(22)}_{1kk}(\lambda)\right)\hat{d}^{(1)}_k(\lambda) = 0,$$

and analogously for $\hat{c}^{(0)}_k(\lambda)$, $\hat{c}^{(1)}_k(\lambda)$:

$$\left(\hat{K}^{(21)}_{2jk}(\lambda) - \hat{K}^{(21)}_{2kk}(\lambda)\right)\hat{c}^{(0)}_k(\lambda) + \left(\hat{K}^{(22)}_{2jk}(\lambda) - \hat{K}^{(22)}_{2kk}(\lambda)\right)\hat{c}^{(1)}_k(\lambda) = 0.$$

We introduce the matrix $E_{jk}(\lambda)$:

$$\begin{pmatrix} \hat{K}^{(11)}_{2jk}(\lambda) - \hat{K}^{(11)}_{2kk}(\lambda) & \hat{K}^{(12)}_{2jk}(\lambda) - \hat{K}^{(12)}_{2kk}(\lambda) & 0 & 0 \\ \hat{K}^{(21)}_{2jk}(\lambda) - \hat{K}^{(21)}_{2kk}(\lambda) & \hat{K}^{(22)}_{2jk}(\lambda) - \hat{K}^{(22)}_{2kk}(\lambda) & 0 & 0 \\ 0 & 0 & \hat{K}^{(11)}_{1jk}(\lambda) - \hat{K}^{(11)}_{1kk}(\lambda) & \hat{K}^{(12)}_{1jk}(\lambda) - \hat{K}^{(12)}_{1kk}(\lambda) \\ 0 & 0 & \hat{K}^{(21)}_{1jk}(\lambda) - \hat{K}^{(21)}_{1kk}(\lambda) & \hat{K}^{(22)}_{1jk}(\lambda) - \hat{K}^{(22)}_{1kk}(\lambda) \end{pmatrix}.$$

Then conditions (9.2.9), (9.2.10) can be formulated as follows:

$$\begin{aligned} &E_{jk}(\lambda)\hat{C}_k(\lambda) = 0, \quad \text{or} \\ &E_{jk}(\lambda)A_k^{-1}(\lambda)\hat{H}_k(\lambda) = 0, \quad \forall j,k = 0,1,\ldots,n-1. \end{aligned} \tag{9.2.14}$$

In case $n = 1$ the conditions (9.2.14) are absent.

9.3. SOME RESULTS

To formulate some results exactly we begin our consideration from perturbed equation (8.1.1):

$$PA^{(\varepsilon)}u_+ = f, \tag{9.3.1}$$

assuming the symbol $A^{(\varepsilon)}(\xi) \in C_\alpha$ (for enough small $\varepsilon > 0$), $A^{(\varepsilon)}(\xi)$ admits wave factorization with respect to cone C^a_+, and

$$\lim_{\varepsilon\to 0} A^{(\varepsilon)}(\xi) = A(\xi)$$

for all ξ, $a^2\xi_2^2 - \xi_1^2 \neq 0$, $A(\xi)$ is an infinitely differentiable homogeneous of order α symbol.

Besides if

$$A^{(\varepsilon)}(\xi) = A^{(\varepsilon)}_{\neq}(\xi)A^{(\varepsilon)}_{=}(\xi)$$

is wave factorization in the sense of Definition 5.1, then

$$A(\xi) = A^{(0)}_{\neq}(\xi)A^{(0)}_{=}(\xi)$$

is a homogeneous wave factorization for $A(\xi)$ (see remark 5.1).

All these assumptions we will be suppose fulfilled.

On symbol $A^{(\varepsilon)}_{\neq}(\xi)$ we construct $\left(a^{(\varepsilon)}_{\neq}\right)^{-1}(x_1, x_2)$ and $b^{(\varepsilon)}_1(x_2)$, $b^{(\varepsilon)}_2(x_1)$ assuming there exist

$$b_1(x_2) = \lim_{\varepsilon \to 0} b^{(\varepsilon)}_1(x_2), \quad x_2 \neq 0,$$

$$b_2(x_1) = \lim_{\varepsilon \to 0} b^{(\varepsilon)}_2(x_1), \quad x_1 \neq 0,$$

and

$$b_1(x_2) \neq 0, \quad \forall x_2 \neq 0, \tag{9.3.2}$$

$$b_2(x_1) \neq 0, \quad \forall x_1 \neq 0. \tag{9.3.3}$$

Further, we compose the determinant $\Delta_A(\lambda)$. In case when $\Delta_A(\lambda) \neq 0$ ($\mathrm{Re}\,\lambda = 1/2$) the system (9.1.5) has a unique solution, and we can define $c(x)$, $d(x)$ knowing this solution. Substituting their in general solution (9.1.2) we will define $\tilde{U}_+(x)$, but this solution generally speaking will not belong to class $H_s(C^a_+)$ because in case æ > 0 the pseudodifferential operator $a^{-1}_{\neq}$ is not bounded in space scale H_s, i.e., the symbol has singularity of order æ at origin. In connection with this fact we will introduce weighted H_s - spaces which "annihilate" such singularity and give the possibility to obtain a priori estimate of the solution there.

Let us denote by $H_{s,æ}(\mathbb{R}^m)$ the space of ditributions $u(x)$ for which their Fourier transforms are locally integrable in the Lebesgue sense function $\tilde{u}(\xi)$ such that

$$||u||^2_{s,æ} = \int\limits_{\mathbb{R}^m} |\tilde{u}(\xi)|^2 |\xi|^{2æ}(1 + |\xi|)^{2(s-æ)} d\xi < +\infty.$$

It is natural to call $H_{s,æ}(\mathbb{R}^m)$ weighted H_s –spaces with weight

$$\left(\frac{|\xi|}{1 + |\xi|}\right)^{æ}.$$

In space scale $H_{s,æ}(\mathbb{R}^m)$, pseudodifferential operators with homogeneous symbols have series of properties needed. Let us give some of them.

Theorem 9.3.1. *Let A be a pseudodifferential operator with symbol which is belonging to $C^\infty_æ$ and homogeneous of order æ. Then*

$$||Au||_{s-æ} \leq c_{s,æ} ||u||_{s,æ}, \quad \forall u \in S(\mathbb{R}^m).$$

Proof. We have

$$||Au||^2_{s-æ} = \int_{\mathbb{R}^m} |A(\xi)u(\xi)|^2(1+|\xi|)^{2(s-æ)}d\xi,$$

and since $A(\xi)$ is homogeneous of order æ and $A(\xi) \in C^\infty \setminus \{0\}$, then

$$c_1 \le \left|A(\xi)|\xi|^{-æ}\right| \le c_2.$$

Hence,

$$||Au||^2_{s-æ} \le c_{s,æ} \int_{\mathbb{R}^m} |\tilde{u}(\xi)|^2|\xi|^{2æ}(1+|\xi|)^{2(s-æ)}d\xi =$$

$$= c_{s,æ}||u||^2_{s,æ}. \quad \blacksquare$$

As before we denote by $[v]_{s,æ}$ the norm in space $H_{s,æ}(\mathbb{R}^{m-1})$

$$[v]^2_{s,æ} = \int_{\mathbb{R}^m} |\tilde{v}(\xi')|^2|\xi'|^{2æ}(1+|\xi'|)^{2(s-æ)}d\xi',$$

$$\xi' = (\xi_1, \dots, \xi_{m-1}).$$

Proof of the following lemma entirely copies the proof of Proposition 3.1.2.

Lemma 9.3.1. *Let $s > 1/2$. Then any function $u(x', x_m) \in H_{s,æ}(\mathbb{R}^m)$ is continuous on $x_m \in \mathbb{R}$ with its values in $H_{s-1/2,æ}(\mathbb{R}^{m-1})$.*

The estimate holds:

$$\max_{x_m\in\mathbb{R}} [u(x', x_m)]_{s-1/2,æ} \le c||u||_{s,æ}, \quad \forall u \in H_{s,æ}(\mathbb{R}^m).$$

Lemma 9.3.2. *Let $c(x_1) \in H_{s-\alpha+1/2}(\mathbb{R}_+)$, A is a pseudodifferential operator with symbol $A_{\neq}(\xi) \in C^\infty\left(\mathbb{R}^2 \setminus \{0\}\right)$ which is homogeneous of order α. Then*

$$||c(x_1)a^{-1}_{\neq}(x)||^2_{s,\alpha} \le c'[c]_{s-\alpha+1/2} \quad (-1 < s-\alpha+1/2 < 0).$$

Proof. We have

$$||c(x_1)a^{-1}_{\neq}(x)||^2_{s,\alpha} = \int_{\mathbb{R}^2} \frac{|\tilde{c}(x_1)|^2}{|a_{\neq}(x)|^2}|x|^{2\alpha}(1+|x|)^{2(s-\alpha)}dx \le$$

$$\le \int_{\mathbb{R}^2} |\tilde{c}(x_1)|^2(1+|x|)^{2(s-\alpha)}dx_1dx_2 \le$$

$$\le c' \int_{-\infty}^{+\infty} |\tilde{c}(x_1)|^2(1+|x_1|)^{2(s-\alpha+1)}dx_1 \int_{-\infty}^{+\infty} (1+|x|)^{-2}dx_2 =$$

$$= c' \int_{-\infty}^{+\infty} |\tilde{c}(x_1)|^2 (1+|x_1|)^{2(s-\alpha+1/2)} dx_1 = c'[c]_{s-\alpha+1/2}. \quad \blacksquare$$

Lemma 9.3.3. *Let $b(x) \in C^\infty(\mathbb{R} \setminus \{0\})$, $b(x)$ be homogeneous of order $1-æ$, and represent symbol of pseudodifferential operator b. Then*

$$||b^{-1}v||_{s-æ+1/2} \le c||v||_{s-1/2,æ-1}, \quad \forall v \in H_{s-1/2,æ-1}(\mathbb{R}).$$

Proof.

$$||b^{-1}v||^2_{s-æ+1/2} = \int_{-\infty}^{+\infty} |b^{-1}(x)|^2 |\tilde{v}(x)|^2 (1+|x|)^{2(s-æ+1/2)} dx \le$$

$$\le c \int_{-\infty}^{+\infty} |x|^{-2(1-æ)} |\tilde{v}(x)|^2 (1+|x|)^{2(s-æ+1/2)} dx =$$

$$= c \int_{-\infty}^{+\infty} \left(\frac{|x|}{1+|x|}\right)^{2(æ-1)} |\tilde{v}(x)|(1+|x|)^{2(s-1/2)} dx =$$

$$= c||v||_{s-1/2,æ-1}. \quad \blacksquare$$

(Statement of Lemma 9.3.3 can be reformulated as following because $s - æ + 1/2 = s - 1/2 - (æ-1)$: if $b(x) \in C^\infty(\mathbb{R} \setminus \{0\})$ and $b(x)$ is homogeneous of order α then

$$||bv||_{s-\alpha} \le c||v||_{s,\alpha},$$

and it follows from Theorem 9.3.1).

Let us note also that

$$||u||_{s,æ} \le c||u||_{s,æ-1}. \tag{9.3.4}$$

In fact,

$$||u||^2_{s,æ} = \int_{\mathbb{R}^m} |\tilde{u}(\xi)|^2 \left(\frac{|\xi|}{1+|\xi|}\right)^{2æ} (1+|\xi|)^{2s} d\xi \le$$

$$\le c \int_{\mathbb{R}^m} |\tilde{u}(\xi)|^2 \left(\frac{|\xi|}{1+|\xi|}\right)^{2(æ-1)} (1+|\xi|)^{2s} d\xi = c||u||_{s,æ-1},$$

because $\left(\frac{|\xi|}{1+|\xi|}\right)^{-1} \ge 1$.

Now we are ready to formulate the result on solvability of the Dirichlet problem.

Theorem 9.3.2. *Let $v_1, v_2 \in H_{s-1/2,æ-1}(\mathbb{R}_+)$, A be an elliptic pseudodifferential operator with symbol $A(\xi) \in C^\infty(\mathbb{R}^2 \setminus \{0\})$ which is homogeneous of order α, æ be an index of wave factorization for $A(\xi)$ with respect to Γ^+. Let the conditions (9.3.2), (9.3.3) be fulfilled, and*

$$\inf |\Delta_A(\lambda)| \ne 0, \quad \operatorname{Re}\lambda = 1/2.$$

Then there exists a unique solution of the Dirichlet problem (9.1.6), (9.1.7) in space $H_{s,æ}(\Gamma^+)$, $æ - s = 1 + \delta$, $|\delta| < 1/2$.

A priori estimate holds

$$||u_+||_{s,æ} \leq c \left([v_1]_{s-1/2,\,æ-1} + [v_2]_{s-1/2,\,æ-1}\right).$$

Proof. Let us consider perturbed equation (9.3.1) under assumptions above (see begining of Section 9.3), and passing to the limit under $\varepsilon \to 0$ we have proved essentially an a priori estimate only.

On Lemma 9.3.2 we obtain (see formula (9.1.2))

$$||u_+||_{s,æ} = ||U_+||_{s,æ} = ||\tilde{U}_+||_{s,æ} \leq$$

$$\leq c_1 \left([c]_{s-æ+1/2} + [d]_{s-æ+1/2}\right),$$

and it is left to estimate $[c]_{s-æ+1/2}$, $[d]_{s-æ+1/2}$ by norms of functions v_1, v_2.

Before let's note the following. At first, $s - æ + 1/2 = -1 - \delta + 1/2 = -1/2 - \delta$, and hence, $-1 < s - æ + 1/2 < 0$, and it gives

$$[c]_{s-æ+1/2} \leq [c]_0 \text{ (analogously for d).} \tag{9.3.5}$$

Second, if $u_+ \in H_{s,æ}(\Gamma^+)$ then v_1, v_2 as traces of this function belong to space $H_{s-1/2,æ}(\mathbb{R}_+)$ in Lemma 9.3.1, but according to statement of the theorem they are taken from class $H_{s-1/2,æ-1}(\mathbb{R}_+)$. This "contradiction" is eliminated by inequality (9.3.4).

After corresponding transformations (see Section 9.1) we obtain the system (9.1.5). Let us denote the matrix of this system by $\hat{A}_4(\lambda)$, and its inverse matrix will be denoted by $\hat{A}_4^{-1}(\lambda)$, if $\Delta_A(\lambda) \neq 0$, $\operatorname{Re}\lambda = 1/2$. Vector with components $\hat{c}_0(\lambda)$, $\hat{c}_1(\lambda)$, $\hat{d}_0(\lambda)$, $\hat{d}_1(\lambda)$ we'll denote $\hat{C}(\lambda)$, and vector with components $\hat{v}_{10}(\lambda)$, $\hat{v}_{11}(\lambda)$, $\hat{v}_{20}(\lambda)$, $\hat{v}_{21}(\lambda)$ we'll denote $\hat{V}(\lambda)$.

Then

$$\hat{C}(\lambda) = \hat{A}_4^{-1}(\lambda)\hat{V}(\lambda),$$

and since components of matrix $\hat{A}_4^{-1}(\lambda)$ are bounded (by virtue of Mellin transform properties [249]), then

$$||\hat{C}(\lambda)||_0 \leq c'||\hat{V}(\lambda)||_0,$$

where

$$||\hat{C}(\lambda)||_0^2 = \frac{1}{2\pi i} \int\limits_{1/2-i\infty}^{1/2+i\infty} |\hat{C}(\lambda)|^2 d\lambda.$$

Now by virtue of Parceval equality

$$\int\limits_0^{+\infty} |C(x)|^2 dx = \frac{1}{2\pi i} \int\limits_{1/2-i\infty}^{1/2+i\infty} |\hat{C}(\lambda)|^2 d\lambda$$

we obtain

$$\int_0^{+\infty} |C(x)|^2 dx \leq c' \int_0^{+\infty} |V(x)|^2 dx,$$

or returning to $c(x)$, $d(x)$,

$$[c]_0 \leq c_1 \left(\left[b_1^{-1} v_1\right]_0 + \left[b_2^{-1} v_2\right]_0 \right),$$

$$[d]_0 \leq c_2 \left(\left[b_1^{-1} v_1\right]_0 + \left[b_2^{-1} v_2\right]_0 \right).$$

By Lemma 9.3.3 we have

$$\left[b_1^{-1} v_1\right]_0 \leq c_1 \left[v_1\right]_{s-1/2, æ-1},$$

$$\left[b_2^{-1} v_2\right]_0 \leq c_2 \left[v_2\right]_{s-1/2, æ-1},$$

from this, taking into account the equality (9.3.5), we obtain

$$[c]_{s-æ+1/2} \leq c_1 \left([v_1]_{s-1/2,æ-1} + [v_2]_{s-1/2,æ-1} \right),$$

$$[d]_{s-æ+1/2} \leq c_2 \left([v_1]_{s-1/2,æ-1} + [v_2]_{s-1/2,æ-1} \right),$$

and then

$$||u_+||_{s,æ} \leq c' \left([v_1]_{s-1/2,æ-1} + [v_2]_{s-1/2,æ-1} \right). \quad \blacksquare$$

Analogous result can be formulated for the Neumann problem and for the Dirichlet-Neumann problem with boundary conditions (9.1.8).

Remark 9.3.1. Theorem 9.3.2 can be formulated in a more general form, namely instead of condition $\Delta_A(\lambda) \neq 0$, $\operatorname{Re} \lambda = 1/2$ requiring solvability of linear integral equations system (9.1.3) and calling it the Shapiro-Lopatinskii condition for angle. In this form, Theorem 9.3.2 can be generalized on a wide class of boundary value problems which are considered in Section 9.2.

9.4. THE CAUCHY PROBLEM

In this section we would like to discuss an interesting phenomenon. The point is the following (we restrict ourselves to a very simple case).

We take a situation when $A(\xi) \in C^\infty(\mathbb{R}^2 \setminus \{0\})$, $A(\xi)$ is homogeneous of order α, $f \equiv 0$, $n = 1$, $a = 1$ considering the equation 9.1.1. Since $n = 1$, we have given one boundary condition, and in Section 9.1 we gave one boundary condition on every angle side. It turns out that such a statement of boundary value problem has sense: giving two boundary conditions on one angle side, but the second side leaving fully free! Let us discuss this more explicitly considering that on the same angle side the Dirichlet and Neumann conditions are given.

So, one considers the problem:

$$(Au_+)(x) = 0, \quad x \in C_+^1,$$
$$u_+\Big|_{\substack{x_2 - x_1 = 0\\ x_1 > 0}} = g_1, \quad \frac{\partial u_+}{\partial n}\Big|_{\substack{x_2 - x_1 = 0\\ x_1 > 0}} = g_2, \tag{9.4.1}$$

where n is normal (inner) to straight line $x_2 - x_1 = 0$.

Acting on scheme Section 9.1, taking into account the structure of general solution (9.1.2), we obtain the following system of linear integral equations for finding of functions $\tilde{c}_0$, $\tilde{d}_0$:

$$\begin{cases} \int\limits_{-\infty}^{+\infty} K(t_1, t_2)c(t_1)dt_1 + d(t_2) = \tilde{r}_1(t_2) \\ \int\limits_{-\infty}^{+\infty} M(t_1, t_2)c(t_1)dt_1 + d(t_2) = \tilde{r}_2(t_2), \end{cases} \tag{9.4.2}$$

where $K(t_1, t_2) = a_{\neq}^{-1}(t_1, t_2)b_1^{-1}(t_2)$, $\tilde{r}_1(t_2) = \tilde{g}_1(t_2) \cdot b_1^{-1}(t_2)$,

$$b_1(t_2) = \int\limits_{-\infty}^{+\infty} a_{\neq}^{-1}(t_1, t_2)dt_1, \quad M(t_1, t_2) = t_1 a_{\neq}^{-1}(t_1, t_2)b_2^{-1}(t_2),$$

$$\tilde{r}_2(t_2) = \tilde{g}_2(t_2)b_2^{-1}(t_2), \quad b_2(t_2) = \int\limits_{-\infty}^{+\infty} t_1 a_{\neq}^{-1}(t_1, t_2)dt_1,$$

assuming that integrals needed exist, and

$$b_i(t_2) \neq 0, \quad \forall t_2 \neq 0, \ i = 1, 2.$$

(it is the Shapiro-Lopatinskii condition on angle side for Dirichlet and Neumann data respectively).

Keeping in mind homogeneity of order -1 for kernels K and M, we apply to system (9.4.2) the procedure above, reducing it to a 4×4 - system of linear integral equations on a half- axis. The determinant of the last system (after applying the Mellin transform) will have the form:

$$\Delta_A(\lambda) = \begin{vmatrix} \hat{K}_{11}(\lambda) & \hat{K}_{12}(\lambda) & 1 & 0 \\ \hat{K}_{22}(\lambda) & \hat{K}_{21}(\lambda) & 0 & 1 \\ \hat{M}_{11}(\lambda) & \hat{M}_{22}(\lambda) & 1 & 0 \\ \hat{M}_{12}(\lambda) & \hat{M}_{21}(\lambda) & 0 & 1 \end{vmatrix},$$

and in the case when it is "separated" from zero (at least, directly one can't see that it is vanishing), we will obtain unique solution for system (9.4.2), and together with this fact we will obtain a unique solution of problem (9.4.1) which is explicitly the Cauchy problem.

10. The Laplacian in a plane infinite angle

10.1. THE DIRICHLET DATA

This section is a "realization" of methods, which are developed in Chapter 9 applied to the Laplacian and contains concrete calculations. Similar [238,239] and more complicated problems [94,134,207,208] were considered earlier by other methods, and as a result, it gave the possibility to obtain the theorem of existence and uniqueness of solution under some restrictions on parameters of functional spaces, size of angle and so on. In this section we have chosen similar functional spaces in which our methods give the possibility to formulate the conditions under fulfilling of which the solution of the posed problem exists and is unique (including explicit construction for solution in terms of Fourier and Mellin transforms). Other approaches one can find in papers which are contained in the list of references from [135].

So, the statement of the problem: finding the function $u_+ \in H_s(C_+^1)$ such that (for simplicity we put $a = 1$)

$$(\Delta u_+)(x) = 0, \quad x \in C_+^1, \tag{10.1.1}$$

and which satisfies the boundary conditions

$$u_+\Big|_{\substack{x_2 = x_1 \\ x_1 > 0}} = g_1(x_2 + x_1), \quad u_+\Big|_{\substack{x_2 = -x_1 \\ x_1 < 0}} = g_1(x_2 - x_1). \tag{10.1.2}$$

In order to do this section independent of previous results we develop our consideration more explicitly than is necessary.

Passing to Fourier images and changing of variables

$$\begin{cases} x_1' = x_1 - x_2 \\ x_2' = x_1 + x_2 \end{cases} \tag{10.1.3}$$

in the formula for general solution (9.1.2) we obtain

$$\tilde{U}_+(t_1, t_2) = a_+^{-1}(t_1, t_2)(c(t_1) + d(t_2)), \tag{10.1.4}$$

(see formula (8.1.5)) where

$$\tilde{U}_+(t_1, t_2) = \tilde{u}_+\left(\frac{t_2 + t_1}{2}, \frac{t_2 - t_1}{2}\right),$$

$$a_{\neq}^{-1}(t_1, t_2) = A_{\neq}^{-1}\left(\frac{t_2 + t_1}{2}, \frac{t_2 - t_1}{2}\right),$$

$$c(t_1) \equiv \tilde{c}_0(t_1), \quad d(t_2) \equiv \tilde{d}_0(t_2).$$

Let us recall that homogeneous wave factorization for the Laplacian is such that (see Chapter 5)

$$A_{\neq}(\xi_1, \xi_2) = \sqrt{2}\xi_2 + \sqrt{\xi_2^2 - \xi_1^2},$$

and its symbol is $\xi_1^2 + \xi_2^2$.

Taking into account the change of variables (10.1.3) one can rewrite the conditions (10.1.2) in the following form:

$$U_+\Big|_{\substack{x_1'=0\\ x_2'>0}} = g_1'(x_2'), \quad U_+\Big|_{\substack{x_2'=0\\ x_1'<0}} = g_2'(x_1'). \tag{10.1.2'}$$

Fourier transform will convert the conditions (10.1.2′) into the followings:

$$\begin{aligned} &\int_{-\infty}^{+\infty} \tilde{U}_+(t_1,t_2)dt_1 = \tilde{g}_1'(t_2),\\ &\int_{-\infty}^{+\infty} \tilde{U}_+(t_1,t_2)dt_2 = \tilde{g}_2'(t_1). \end{aligned} \tag{10.1.2''}$$

We recall once again the formal scheme for solving of boundary value problem (10.1.1), (10.1.2). Integrating two parts of (10.1.4) at first on t_1, and then on t_2, we have (taking into account (10.1.2))

$$\int_{-\infty}^{+\infty} a_{\neq}^{-1}(t_1,t_2)c(t_1)dt_1 + d(t_2)\int_{-\infty}^{+\infty} a_{\neq}^{-1}(t_1,t_2)dt_1 = \tilde{g}_1'(t_2), \tag{10.1.5}$$

$$c(t_1)\int_{-\infty}^{+\infty} a_{\neq}^{-1}(t_1,t_2)dt_2 + \int_{-\infty}^{+\infty} a_{\neq}^{-1}(t_1,t_2)d(t_2)dt_2 = \tilde{g}_2'(t_1). \tag{10.1.6}$$

Since the symbol $A_{\neq}(t_1,t_2)$ has concrete form, let us write it in t_1, t_2 - variables:

$$a_{\neq}(t_1,t_2) = \begin{cases} \dfrac{t_2-t_1}{\sqrt{2}} + \sqrt{-t_1t_2}, & t_1t_2 < 0, \quad t_2 > t_1\\ \dfrac{t_2-t_1}{\sqrt{2}} - \sqrt{-t_1t_2}, & t_1t_2 < 0, \quad t_2 < t_1\\ \dfrac{t_2-t_1}{\sqrt{2}} + i\sqrt{t_1t_2}, & t_1t_2 > 0, \end{cases}$$

and, hence, the integrals

$$b_1(t_2) = \int_{-\infty}^{+\infty} a_{\neq}^{-1}(t_1,t_2)dt_1, \quad b_2(t_1) = \int_{-\infty}^{+\infty} a_{\neq}^{-1}(t_1,t_2)dt_2,$$

must be calculated.

Let us make a small reservation. To use the Theorem 9.3.2 (and respectively, the Definition 5.1) we must consider the symbol $\xi_1^2 + \xi_2^2 + \varepsilon^2$, factorize it as in Example 5.1, develop previous calculations, and then pass to the limit under $\varepsilon \to 0$.

This passage to the limit we realize now when we write out the equations (10.1.5), (10.1.6).

Since $a_{\neq}(t_1, t_2)$ is homogeneous of order 1, then it is not difficult to satisfy oneself that both functions $b_1(t_2)$ and $b_2(t_1)$ will be homogeneous of order 0, and hence, it is enough to calculate $b_1(\pm 1)$, $b_2(\pm 1)$.

Let us consider

$$b_1(1) = \int_{-\infty}^{+\infty} \frac{dt}{a_{\neq}(t,1)},$$

and write

$$a_{\neq}(t,1) = \begin{cases} \dfrac{1-t}{\sqrt{2}} + \sqrt{-t}, & t < 0, \quad t < 1 \\ \dfrac{1-t}{\sqrt{2}} - \sqrt{-t}, & t < 0, \quad t > 1 \\ \dfrac{1-t}{\sqrt{2}} + i\sqrt{t}, & t > 0 \ . \end{cases}$$

The second situation for $a_{\neq}(t,1)$ is impossible, and we will represent

$$a_{\neq}(t,1) = \frac{1-t}{\sqrt{2}} + \sqrt{-t}$$

for all $t \in \mathbb{R}$ taking such branch $\sqrt{z}$ of square root which maps onto the upper half-plane.

$$\int_{-\infty}^{+\infty} \frac{e^{-ixt}dt}{a_{\neq}(t,1)} = \sqrt{2} \int_{-\infty}^{+\infty} \frac{e^{-ixt}dt}{1-t+\sqrt{-2t}} =$$

$$(\text{change } -\text{t} = \text{y})$$

$$= \sqrt{2} \int_{-\infty}^{+\infty} \frac{e^{ixy}dt}{y+\sqrt{2y}+1}.$$

Decomposing the the denominator on "simple" fractions we have

$$\int_{-\infty}^{+\infty} \frac{e^{-ixt}dt}{a_{\neq}(t,1)} = -i\left(\int_{-\infty}^{+\infty} \frac{e^{ixy}dy}{\sqrt{y}+\frac{1}{\sqrt{2}}(1-i)} - \int_{-\infty}^{+\infty} \frac{e^{ixy}dy}{\sqrt{y}+\frac{1}{\sqrt{2}}(1+i)} \right). \tag{10.1.7}$$

Since in the second integral the function standing under the integral sign is analytical in the upper and lower half-plane, it is equal to zero by the residue theorem.

Let us note that roots of trinomial $y+\sqrt{2y}+1$ are equal to

$$(\sqrt{y})_{1;2} = -\frac{1}{\sqrt{2}}(1 \mp i).$$

By virtue of choice of branch we have only one root, it is $e^{i\frac{3\pi}{4}}$, and to it corresponds the y equal to $-i$.

Then the first integral in formula (10.1.7) under $x > 0$ is equal to 0 by the residue theorem, but for $x < 0$

$$-i \int_{-\infty}^{+\infty} \frac{e^{ixy}\,dy}{\sqrt{y} + \frac{1}{\sqrt{2}}(1-i)} = -2\pi i \operatorname*{Res}_{z=-i} \left[\frac{e^{ixz}}{\sqrt{z} + \frac{1}{\sqrt{2}}(1-i)} \right] =$$

$$= -4\pi i e^{x} \sqrt{-i} = 4\pi i e^{-i\frac{\pi}{4}} e^{x},$$

and it implies

$$\int_{-\infty}^{+\infty} \frac{dt}{a_{\neq}(t,1)} = -4\pi e^{-i\frac{\pi}{4}}.$$

Reasoning analogously we'll find

$$\int_{-\infty}^{+\infty} \frac{dt}{a_{\neq}(t,-1)} = 4\pi e^{i\frac{\pi}{4}}.$$

So,

$$b_1(\pm 1) = \mp e^{\mp i\frac{\pi}{4}}.$$

Turn to calculation of $b_2(\pm 1)$. Let's write

$$a_{\neq}(1,t) = \begin{cases} \dfrac{t-1}{\sqrt{2}} + \sqrt{-t}, & t < 0, \quad t > 1 \\ \dfrac{t-1}{\sqrt{2}} - \sqrt{-t}, & t < 0, \quad t < 1 \\ \dfrac{t-1}{\sqrt{2}} + i\sqrt{t}, & t > 0 \end{cases}.$$

The first situation does not occur, and we will represent

$$a_{\neq}(1,t) = \frac{t-1}{\sqrt{2}} - \sqrt{-t}$$

for all $t \in \mathbb{R}$ taking such branch $\sqrt{z}$ which maps onto the lower half-plane.

$$\int_{-\infty}^{+\infty} \frac{e^{-ixt}\,dt}{a_{\neq}(1,t)} = \sqrt{2} \int_{-\infty}^{+\infty} \frac{e^{-ixt}\,dt}{t - \sqrt{-2t} - 1} = -\sqrt{2} \int_{-\infty}^{+\infty} \frac{e^{ixt}\,dt}{t + \sqrt{2t} + 1} =$$

$$= i \left(\int_{-\infty}^{+\infty} \frac{e^{ixt}\,dt}{\sqrt{t} + \frac{1}{\sqrt{2}}(1-i)} - \int_{-\infty}^{+\infty} \frac{e^{ixt}\,dt}{\sqrt{t} + \frac{1}{\sqrt{2}}(1+i)} \right).$$

The function under the first integral sign is analytical in the upper and lower half-plane, and hence this is equal to zero.

The second integral under $x < 0$ is equal to 0 by the residue theorem, but under $x > 0$

$$-i\int_{-\infty}^{+\infty}\frac{e^{ixt}dt}{\sqrt{t}+\frac{1}{\sqrt{2}}(1+i)} = 2\pi i \operatorname*{Res}_{z=i}\left[\frac{e^{ixz}}{\sqrt{z}+\frac{1}{\sqrt{2}}(1+i)}\right] =$$

$$= 4\pi i e^{-x}\sqrt{i} = -4\pi i e^{-x} e^{i\frac{\pi}{4}},$$

and it implies

$$\int_{-\infty}^{+\infty}\frac{dt}{a_{\neq}(1,t)} = 4\pi e^{i\frac{\pi}{4}}.$$

Analogously the last left integral is calculated, and we have

$$b_2(\pm 1) = \pm 4\pi e^{\pm i\frac{\pi}{4}}.$$

Further we work on the scheme from Section 9.1. Whenever $b_1(t_2) \neq 0$, $\forall t_2 \neq 0$, $b_2(t_1) \neq 0$, $\forall t_1 \neq 0$, we divide two parts of equation (10.1.5), (10.1.6) by $b_1(t_2), b_2(t_1)$ respectively, and introduce the following notations beforehand writing (10.1.5), (10.1.6) in the form of system

$$\begin{cases} \int_{-\infty}^{+\infty} K(t_1,t_2)c(t_1)dt_1 + d(t_2) = \tilde{r}_1(t_2) \\ c(t_1) + \int_{-\infty}^{+\infty} M(t_1,t_2)d(t_2)dt_2 = \tilde{r}_2(t_1) \end{cases} \tag{10.1.8}$$

where

$$K(t_1,t_2) = a_{\neq}^{-1}(t_1,t_2)b_1^{-1}(t_2), \quad M(t_1,t_2) = a_{\neq}^{-1}(t_1,t_2)b_2^{-1}(t_1),$$

$$\tilde{r}_1(t_2) = \tilde{g}_1'(t_2)b_1^{-1}(t_2), \quad \tilde{r}_2(t_1) = \tilde{g}_2'(t_1)b_2^{-1}(t_1).$$

As before in Section 9.1 for all $t_1 > 0$, $t_2 > 0$ let us denote:

$$K_{11}(t_1,t_2) = K(t_1,t_2), \quad K_{12}(t_1,t_2) = K(-t_1,t_2),$$

$$K_{21}(t_1,t_2) = K(-t_1,-t_2), \quad K_{22}(t_1,t_2) = K(t_1,-t_2),$$

and analogously we define the kernels $M_{ij}(t_1,t_2)$, $i,j = 1,2$.

Further we denote $c_0(t_1)$ the restriction $c(t_1)$ on $(0;+\infty)$, and $c_1(t_1)$ is the restriction $c(t_1)$ on $(-\infty;0)$ and analogously for $d_0(t_2)$, $d_1(t_2)$. Finally, we define $r_{10}(t_2)$ as restriction $\tilde{r}_1(t_2)$ on $(0;+\infty)$ and $r_{11}(t_2)$ as restriction $\tilde{r}_1(-t_2)$ on $(0;+\infty)$ and by the same way for $r_{20}(t_1)$, $r_{21}(t_2)$. A system of two linear integral equations (10.1.8) on the whole axis with respect to two unknown functions $c(t_1)$, $d(t_2)$ will be written in the form of a system of four linear integral equations with respect to four unknown functions $c_0(t_1)$, $c_1(t_1)$, $d_0(t_2)$, $d_1(t_2)$ on the positive half-axis, and the kernels of this system will be homogeneous of order -1, it can be studied with help of the Mellin transform. Determinant of this system (more precisely, already of system of linear algebraic equations with respect to unknowns $\hat{c}_0(\lambda)$, $\hat{c}_1(\lambda)$, $\hat{d}_0(\lambda)$, $\hat{d}_1(\lambda)$ after applying of Mellin transform) must have the following form in which we denote $\hat{K}_{ij}(\lambda)$, $\hat{M}_{ij}(\lambda)$ the Mellin transforms of functions $K_{ij}(1,t)$, $M_{ij}(t,1)$, $i,j=1,2$, respectively:

$$\Delta_A(\lambda) = \begin{vmatrix} \hat{K}_{11}(\lambda) & \hat{K}_{12}(\lambda) & 1 & 0 \\ \hat{K}_{22}(\lambda) & \hat{K}_{21}(\lambda) & 0 & 1 \\ 1 & 0 & \hat{M}_{11}(\lambda) & \hat{M}_{22}(\lambda) \\ 0 & 1 & \hat{M}_{12}(\lambda) & \hat{M}_{21}(\lambda) \end{vmatrix}.$$

Now it is necessary to calculate the Mellin transform of functions $K_{ij}(1,t)$, $M_{ij}(t,1)$, $i,j = 1,2$. Since in this situation $b_1(t_2)$, $b_2(t_1)$ will play the role of constants we temporarily omit them, and we will consider them when we compose the determinant $\Delta_A(\lambda)$. We have

$$\underline{K}_{11}^{-1}(1,t) = a_{\neq}(1,t) = \frac{t-1}{\sqrt{2}} + i\sqrt{t},$$

$$\underline{K}_{12}^{-1}(1,t) = a_{\neq}(-1,t) = \frac{t+1}{\sqrt{2}} + \sqrt{t},$$

$$\underline{K}_{21}^{-1}(1,t) = a_{\neq}(-1,-t) = \frac{1-t}{\sqrt{2}} + i\sqrt{t},$$

$$\underline{K}_{22}^{-1}(1,t) = a_{\neq}(1,-t) = -\left(\frac{t+1}{\sqrt{2}} + \sqrt{t}\right).$$

Analogously we calculate $M_{ij}(t,1)$:

$$\underline{M}_{11}^{-1}(t,1) = a_{\neq}(t,1) = a_{\neq}(-1,-t) = \underline{K}_{21}^{-1}(1,t),$$

$$\underline{M}_{12}^{-1}(t,1) = a_{\neq}(t,-1) = a_{\neq}(1,-t) = \underline{K}_{22}^{-1}(1,t),$$

$$\underline{M}_{21}^{-1}(t,1) = a_{\neq}(-t,-1) = a_{\neq}(1,t) = \underline{K}_{11}^{-1}(1,t),$$

$$\underline{M}_{22}^{-1}(t,1) = a_{\neq}(-t,1) = a_{\neq}(-1,t) = \underline{K}_{12}^{-1}(1,t).$$

Hence, it is enough to calculate four Mellin transforms.

We will use the well-known Mellin transform from [66]: the Mellin transform of function

$$(1+at)^{-n-1}, \quad |\arg a| < \pi, \tag{10.1.8'}$$

is the function

$$(-1)^n \frac{\pi}{a^\lambda} \operatorname{cosec}(\pi\lambda) \binom{\lambda-1}{n}, \quad 0 < \operatorname{Re}\lambda < n+1, \tag{10.1.8''}$$

where $\binom{\lambda-1}{n}$ denotes the expression $\frac{\Gamma(\lambda)}{\Gamma(n+1)\Gamma(\lambda-n)}$, Γ is the Euler gamma-function.

(Immediately note that in our case $n=0$ we have $\binom{\lambda-1}{0} \equiv 1$).

Let us write

$$\hat{\underline{K}}_{11}(\lambda) = \int_0^{+\infty} K_{11}(1,t)t^{\lambda-1}dt = \sqrt{2}\int_0^{+\infty} \frac{t^{\lambda-1}dt}{t-1+\sqrt{2t}i} =$$

$$= 2\sqrt{2}\int_0^{+\infty} \frac{y^{2\lambda-1}dy}{y^2+\sqrt{2}iy-1}.$$

Finding the roots of trinomial $y^2+\sqrt{2}iy-1$, it is not difficult to verify that the representation holds:

$$\frac{1}{y^2+\sqrt{2}iy-1} = \frac{1}{\sqrt{2}y-1+i} - \frac{1}{\sqrt{2}y+1+i},$$

and it is more convenient for us if its summands are written as

$$\frac{1}{\sqrt{2}y+i-1} = -\frac{i+1}{2}\left(1+\frac{\sqrt{2}}{i-1}y\right)^{-1},$$

$$\frac{1}{\sqrt{2}y+i+1} = -\frac{i-1}{2}\left(1+\frac{\sqrt{2}}{i+1}y\right)^{-1},$$

and from where according to (10.1.8′), (10.1.8″) we obtain

$$\hat{\underline{K}}_{11}(\lambda) = \sqrt{2}(i-1)\frac{\pi}{\left(\frac{\sqrt{2}}{i+1}\right)^{2\lambda}}\operatorname{cosec}(2\pi\lambda) -$$

$$-\sqrt{2}(i+1)\frac{\pi}{\left(\frac{\sqrt{2}}{i-1}\right)^{2\lambda}}\operatorname{cosec}(2\pi\lambda) = \cdots =$$

(after some identical transformations)

$$= -2\pi\operatorname{cosec}(2\pi\lambda)\left(e^{i\frac{\pi}{4}(2\lambda-1)} - e^{i\frac{3\pi}{4}(2\lambda-1)}\right).$$

Acting in the same way we obtain as a result:

$$\hat{\underline{K}}_{12}(\lambda) = -4\pi\operatorname{cosec}(2\pi\lambda)\sin\frac{\pi}{4}(2\lambda-1),$$

$$\hat{\underline{K}}_{22}(\lambda) = 4\pi\operatorname{cosec}(2\pi\lambda)\sin\frac{\pi}{4}(2\lambda-1),$$

$$\hat{\underline{K}}_{21}(\lambda) = -2\pi\operatorname{cosec}(2\pi\lambda)\left(e^{-\frac{3\pi}{4}(2\lambda-1)} - e^{-i\frac{\pi}{4}(2\lambda-1)}\right).$$

Before writing out the determinant $\Delta_A(\lambda)$ we consider some remarks which permit us to simplify our calculations.

Let us consider two 2×2 - matrices

$$A = \begin{pmatrix} a_{11} & a_{12} \\ a_{21} & a_{22} \end{pmatrix}, \quad B = \begin{pmatrix} b_{11} & b_{12} \\ b_{21} & b_{22} \end{pmatrix},$$

I is a unit 2×2 - matrix, and let us compose 4×4 –matrix

$$C = \begin{pmatrix} A & \mathrm{I} \\ \mathrm{I} & B \end{pmatrix}.$$

Lemma 10.1.1. $\det C = \det A \cdot \det B - \mathrm{sp}\,(A \cdot B) + 1$, *where* sp *denotes the trace of the matrix.*

The proof can be obtained by direct calculation.

Let us denote (here already we consider the constants which have been mentioned above)

$$K(\lambda) = \begin{pmatrix} \hat{K}_{11}(\lambda) & \hat{K}_{12}(\lambda) \\ \hat{K}_{22}(\lambda) & \hat{K}_{21}(\lambda) \end{pmatrix}, \quad M(\lambda) = \begin{pmatrix} \hat{M}_{11}(\lambda) & \hat{M}_{22}(\lambda) \\ \hat{M}_{12}(\lambda) & \hat{M}_{21}(\lambda) \end{pmatrix}.$$

Taking into account the relations between $\hat{K}_{ij}(\lambda)$ and $\hat{M}_{ij}(\lambda)$ we have

$$M(\lambda) = \begin{pmatrix} \hat{K}_{21}(\lambda) & \hat{K}_{12}(\lambda) \\ \hat{K}_{22}(\lambda) & \hat{K}_{11}(\lambda) \end{pmatrix},$$

from where we directly obtain that $\det K(\lambda) = \det M(\lambda)$.

Besides one can note that

$$\hat{K}_{12}^2(\lambda) = -\hat{K}_{22}^2(\lambda),$$

and, hence, the following result is valid.

Lemma 10.1.2.$\Delta_A(\lambda) = \det^2 K(\lambda) - 2\hat{K}_{11}(\lambda)\hat{K}_{21}(\lambda) + 1.$

With the help of non-complicated considerations one obtains

Lemma 10.1.3. *Under* $\mathrm{Re}\,\lambda = \mathrm{const}$, $\mathrm{Im}\,\lambda \to \pm\infty$ *we have*

$$\hat{K}_{ij}(\lambda) \to 0,$$

and, hence $\Delta_A(\lambda) \to 1$.

The final point in studying of determinant $\Delta_A(\lambda)$ gives

Lemma 10.1.4. $\det K(\lambda) \equiv 0$,

$$\Delta_A(\lambda) = 1 - 2\mathrm{cosec}^2\,(2\pi\lambda)\sin^2\frac{\pi}{4}(2\lambda - 1),$$

and the straight line $\mathrm{Re}\,\lambda = 1/2$ *does not contain* $\Delta_A(\lambda)$ *- zeroes.*

Lemma 10.1.4 permits us to apply our previous results (see Section 9.1) because the conditions required are fulfilled.

To formulate a precise result we introduce some functional spaces with weight.

Let us denote as before $H_{s,\alpha}(\mathbb{R}^2)$ the space of distributions $u(x)$ for which their Fourier transforms are locally integrable in Lebesgue sense function $\tilde{u}(\xi)$ such that

$$||u||_{s,\alpha}^2 = \int\limits_{\mathbb{R}^2} |\tilde{u}(\xi)|^2 |\xi|^{2\alpha}(1 + |\xi|)^{2(s-\alpha)} d\xi < +\infty.$$

As before we will be denote $[v]_{s,\alpha}$ the norm of function v in space $H_{s,a}(\mathbb{R})$:

$$[v]_{s,\alpha}^2 = \int_{-\infty}^{+\infty} |\tilde{v}(t)|^2 |t|^{2\alpha}(1+|t|)^{2(s-\alpha)}dt.$$

By insignificant modification of our arguments (see n° 9.1) one proves the following

Theorem 10.1.1. *Let $g_1, g_2 \in H_s(\mathbb{R}_+)$, $s > 1/2$. Then the problem (10.1.1), (10.1.2) has a unique solution in space $H_{s,1}(C_+^1)$, and a priori estimate holds*

$$||u_+||_{s,1} \leq c\left([g_1]_s + [g_2]_s\right).$$

Let us note that from the statement above follows that this solution can be explicitly written in terms of Fourier and Mellin transforms.

10.2. THE DIRICHLET-NEUMANN DATA

Here we consider the following boundary value problem: finding the function u_+ defined in C_+^1 which satisfies the Laplace equation:

$$(\Delta u_+)(x) = 0, \quad x \in C_+^1, \tag{10.2.1}$$

and on angle sides ∂C_+^1 it satisfies boundary Dirichlet condition (on one side) and boundary Neumann condition (on the other side):

$$u_+\Big|_{\substack{x_2 = -x_1 \\ x_1 < 0}} = g_1, \quad \frac{\partial u_+}{\partial n}\Big|_{\substack{x_2 = x_1 \\ x_1 > 0}} = g_2, \tag{10.2.2}$$

where n is normal to straight line $x_2 = x_1$.

The solution u_+ we will seek in scale $H_{s,æ}(C_+^1)$ of weighted S.L.Sobolev-L.N.Slobodetskii spaces, and since in this space scale pseudodifferential operators have properties which are analogous to their in usual H_s - spaces (see Proposition 3.1.2, 4.1.1 and Theorem 9.3.1, Lemma 9.3.1) then the right sides g_1, g_2 in boundary conditions (10.2.2) must be taken from $H_{s-3/2}(\mathbb{R}_+)$, $H_{s-1/2}(\mathbb{R}_-)$ (perhaps with some weight) respectively.

As before we perturb the Laplacian, hence its symbol will be $\xi_1^2 + \xi_2^2 + \varepsilon^2$ (to apply Definition 5.1), and in final formulas we must put $\varepsilon \to 0$. We do it now obtaining the homogeneous wave factorization for symbol $\xi_1^2 + \xi_2^2$ (see example 5.1).

General solution of equation (10.2.1) has the same form as in Section 10.1 and by change of variables in (10.2.1), (10.2.2) and passing to Fourier images (taking into account the properties of Fourier transform and boundary conditions) instead of our general solution (8.1.5) we obtain

$$\tilde{U}_+(t_1, t_2) = a_{\neq}^{-1}(t_1, t_2)(c(t_1) + d(t_2)), \tag{10.2.3}$$

and two conditions to define unknown functions c, d :

$$\int_{-\infty}^{+\infty} t_1 \tilde{U}_+(t_1,t_2)dt_1 = \tilde{g}_1'(t_2), \qquad \int_{-\infty}^{+\infty} \tilde{U}_+(t_1,t_2)dt_2 = \tilde{g}_2'(t_1). \tag{10.2.4}$$

We preserve here all notations from n° 10.1.

Substituting (10.2.3) in (10.2.4) we obtain the following system of linear integral equations with respect to $c(t_1)$, $d(t_2)$:

$$\begin{cases} \int_{-\infty}^{+\infty} t_1 a_{\neq}^{-1}(t_1,t_2)c(t_1)dt_1 + d(t_2)\int_{-\infty}^{+\infty} t_1 a_{\neq}^{-1}(t_1,t_2)dt_1 = \tilde{g}_1'(t_2) \\ c(t_1)\int_{-\infty}^{+\infty} a_{\neq}^{-1}(t_1,t_2)dt_2 + \int_{-\infty}^{+\infty} a_{\neq}^{-1}(t_1,t_2)d(t_2)dt_2 = \tilde{g}_2'(t_1) \,. \end{cases} \tag{10.2.5}$$

To simplify the system (10.2.5) (taking into account the concrete form of symbol $a_{\neq}(t_1,t_2)$) we calculate the following two integrals:

$$\int_{-\infty}^{+\infty} a_{\neq}^{-1}(t_1,t_2)dt_2, \qquad \int_{-\infty}^{+\infty} t_1 a_{\neq}^{-1}(t_1,t_2)dt_1.$$

Since the function

$$b(t_1) = \int_{-\infty}^{+\infty} a_{\neq}^{-1}(t_1,t_2)dt_2$$

is homogeneous of order 0, then it is enough to calculate $b(\pm 1)$ and it was done in n° 10.1 :

$$b(\pm 1) = \mp 4\pi e^{\pm i\frac{\pi}{4}}.$$

The second integral

$$b_1(t_2) = \int_{-\infty}^{+\infty} \frac{t_1 dt_1}{a_{\neq}(t_1,t_2)}$$

is a function which is homogeneous of order 1, and it will be calculated in the following way: at first we will calculate $b(\pm 1)$ and then

$$b_1(t_2) = \begin{cases} b_1(1)t_2, & t_2 > 0 \\ b_1(-1)(-t_2), & t_2 < 0. \end{cases}$$

Let us write

$$b_1(1) = \int_{-\infty}^{+\infty} \frac{tdt}{a_{\neq}(t,1)} = i\left(\int_{-\infty}^{+\infty} \frac{e^{-ixt}dt}{a_{\neq}(t,1)}\right)_x'\Bigg|_{x=0},$$

and start from calculation of

$$\int_{-\infty}^{+\infty} \frac{e^{-ixt}dt}{a_{\neq}(t,1)}.$$

By definition of function $a_{\neq}(t,1)$ (example 5.1) we have

$$a_{\neq}(t,1) = \begin{cases} \dfrac{1-t}{\sqrt{2}} + \sqrt{-t}, & t<0, \quad t<1 \\ \dfrac{1-t}{\sqrt{2}} - \sqrt{-t}, & t<0, \quad t>1 \\ \dfrac{1-t}{\sqrt{2}} + i\sqrt{t}, & t>0 \end{cases}.$$

Since the second situation for $a_{\neq}(t,1)$ is impossible then we'll represent

$$a_{\neq}(t,1) = \frac{1-t}{\sqrt{2}} + \sqrt{-t}$$

for all $t \in \mathbb{R}$ taking such branch of $\sqrt{z}$ for which its image is the upper half-plane. We have

$$\int_{-\infty}^{+\infty} \frac{e^{-ixt}dt}{a_{\neq}(t,1)} = \sqrt{2}\int_{-\infty}^{+\infty} \frac{e^{-ixt}dt}{1-t+\sqrt{-2t}} = \sqrt{2}\int_{-\infty}^{+\infty} \frac{e^{ixy}dy}{y+\sqrt{2y}+1}.$$

Expanding the denominator on "simple fractions" we obtain

$$\int_{-\infty}^{+\infty} \frac{e^{-ixt}dt}{a_{\neq}(t,1)} = -i\left(\int_{-\infty}^{+\infty} \frac{e^{ixy}dy}{\sqrt{y}+\frac{1}{\sqrt{2}}(1-i)} - \int_{-\infty}^{+\infty} \frac{e^{ixy}dy}{\sqrt{y}+\frac{1}{\sqrt{2}}(1+i)}\right). \tag{10.2.6}$$

In the second integral the function under the integral sign is analytic in the upper and lower half-plane, and hence it is equal to zero by the residue theorem.

Roots of trinomial $y+\sqrt{2y}+1$ are equal to $(\sqrt{y})_{1;2} = -\frac{1}{\sqrt{2}}(1\pm i)$, but by virtue of branch choice the root is unique, it is $e^{-i\frac{3\pi}{4}}$, and $y=-i$ corresponds to this root.

The first integral in formula (10.2.6) under $x>0$ is equal to zero by the residue theorem, but for $x<0$

$$\int_{-\infty}^{+\infty} \frac{e^{ixy}dy}{\sqrt{y}+\frac{1}{\sqrt{2}}(1-i)} = -2\pi i \operatorname*{Res}_{z=-i}\left[\frac{e^{ixz}}{\sqrt{z}+\frac{1}{\sqrt{2}}(1-i)}\right] =$$

$$= -4\pi i e^{x}\sqrt{-i} = 4\pi i e^{-i\frac{\pi}{4}} e^{x},$$

from which follows that

$$\int_{-\infty}^{+\infty} \frac{e^{-ixt}dt}{a_{\neq}(t,1)} = 4\pi e^{x-i\frac{\pi}{4}}.$$

Hence,

$$b_1(1) = 4\pi i e^{-i\frac{\pi}{4}} = 4\pi e^{i\frac{\pi}{4}}.$$

Analogously,

$$a_{\neq}(t,-1) = \begin{cases} \dfrac{-1-t}{\sqrt{2}} + \sqrt{t}, & -t < 0, \quad -1 > t \\ \dfrac{-1-t}{\sqrt{2}} - \sqrt{t}, & -t < 0, \quad -1 < t \\ \dfrac{-1-t}{\sqrt{2}} + i\sqrt{-t}, & -t > 0 \end{cases},$$

or, rewriting

$$a_{\neq}(t,-1) = \begin{cases} \dfrac{-1-t}{\sqrt{2}} + \sqrt{t}, & t > 0, \quad t < -1 \\ \dfrac{-1-t}{\sqrt{2}} - \sqrt{t}, & t > 0, \quad t > -1 \\ \dfrac{-1-t}{\sqrt{2}} + i\sqrt{-t}, & t < 0 \end{cases},$$

and since the first situation is impossible we'll write

$$a_{\neq}(t,-1) = \frac{-1-t}{\sqrt{2}} - \sqrt{t}$$

for all $t \in \mathbb{R}$ taking such branch of $\sqrt{z}$ for which its image is the lower half-plane. We have

$$\int_{-\infty}^{+\infty} \frac{e^{-ixt}dt}{a_{\neq}(t,-1)} = -\sqrt{2}\int_{-\infty}^{+\infty} \frac{e^{-ixt}dt}{t+\sqrt{2t}+1} =$$

$$= i\left(\int_{-\infty}^{+\infty} \frac{e^{-ixt}dt}{\sqrt{t}+\frac{1}{\sqrt{2}}(1-i)} - \int_{-\infty}^{+\infty} \frac{e^{-ixt}dt}{\sqrt{t}+\frac{1}{\sqrt{2}}(1+i)}\right),$$

and arguing further as before we have for the second integral under $x < 0$

$$-i\int_{-\infty}^{+\infty} \frac{e^{-ixt}dt}{\sqrt{t}+\frac{1}{\sqrt{2}}(1+i)} = -i\left(2\pi i \operatorname*{Res}_{z=i}\left[\frac{e^{-ixz}}{\sqrt{z}+\frac{1}{\sqrt{2}}(1+i)}\right]\right) =$$

$$= -i(4\pi i e^{x}\sqrt{i}) = 4\pi e^{x} e^{-\frac{3\pi}{4}i} = -4\pi e^{x} e^{\frac{\pi}{4}i} = -4\pi e^{\left(x+i\frac{\pi}{4}\right)},$$

and then

$$b_1(-1) = -4\pi i e^{i\frac{\pi}{4}} = -4\pi e^{i\frac{3\pi}{4}} = 4\pi e^{-i\frac{\pi}{4}}.$$

Let us return to the system of equation (10.2.5) which we will rewrite in following form:

$$\begin{cases} \displaystyle\int_{-\infty}^{+\infty} K(t_1,t_2)c(t_1)dt_1 + d(t_2) = \tilde{r}_1(t_2) \\ \displaystyle c(t_1) + \int_{-\infty}^{+\infty} M(t_1,t_2)d(t_2)dt_2 = \tilde{r}_2(t_1) \end{cases} \qquad (10.2.7)$$

where the following notations are introduced:

$$K(t_1,t_2)=a_{\neq}^{-1}(t_1,t_2)b_1^{-1}(t_2)t_1,\quad M(t_1,t_2)=a_{\neq}^{-1}(t_1,t_2)b^{-1}(t_1),$$

$$\tilde{r}_1(t_2)=\tilde{g}_1'(t_2)b_1^{-1}(t_2),\quad \tilde{r}_2(t_1)=\tilde{g}_2'(t_1)b^{-1}(t_1).$$

As before (see Section 10.1) we define the $K_{ij}(t_1,t_2)$, $M_{ij}(t_1,t_2)$ $i,j=1,2$, and introduce the functions $c_0(t_1)$, $c_1(t_1)$, $d_0(t_2)$, $d_1(t_2)$ which are defined on the positive half-axis. As a result, instead of a system of two linear integral equations (10.2.7) with respect to two unknown functions $c(t_1)$, $d(t_2)$, we have a system of four linear integral equations with respect to four unknown functions $c_0(t_1)$, $c_1(t_1)$, $d_0(t_2)$, $d_1(t_2)$ on the positive half-axis. The kernels $K_{ij}(t_1,t_2)$ up to constant have the form $t_1t_2^{-1}a_{\neq}^{-1}(t_1,t_2)$, and the kernels $M_{ij}(t_1,t_2)$ have the form $a_{\neq}^{-1}(t_1,t_2)$, and hence, are homogeneous of order -1 functions. The last permits us to apply Mellin transform for study of this system solvability.

The Mellin transform reduces the 4×4 - system of linear integral equations mentioned to a 4×4 - system of linear algebraic equations with respect to unknowns $\hat{c}_0(\lambda)$, $\hat{c}_1(\lambda)$, $\hat{d}_0(\lambda)$, $\hat{d}_1(\lambda)$, where "$\hat{\ }$" denotes the Mellin transform of corresponding function, with determinant

$$\Delta_A(\lambda)=\begin{vmatrix}\hat{K}_{11}(\lambda) & \hat{K}_{12}(\lambda) & 1 & 0\\ \hat{K}_{22}(\lambda) & \hat{K}_{21}(\lambda) & 0 & 1\\ 1 & 0 & \hat{M}_{11}(\lambda) & \hat{M}_{22}(\lambda)\\ 0 & 1 & \hat{M}_{12}(\lambda) & \hat{M}_{21}(\lambda)\end{vmatrix},$$

$\hat{K}_{ij}(\lambda)$, $M_{ij}(\lambda)$ denotes the Mellin transform of functions $K_{ij}(1,t)$, $M_{ij}(t,1)$, $i,j=1,2$, respectively. So, it is necessary to calculate the Mellin transform of functions $K_{ij}(1,t)$, $M_{ij}(t,1)$.

We have

$$K_{11}(1,t)=a_{\neq}^{-1}(1,t)b_1^{-1}(t)=(4\pi)^{-1}e^{-i\frac{\pi}{4}}t^{-1}\left(\frac{t-1}{\sqrt{2}}+i\sqrt{t}\right)^{-1},$$

$$K_{12}(1,t)=-a_{\neq}^{-1}(-1,t)b_1^{-1}(t)=-(4\pi)^{-1}e^{-i\frac{\pi}{4}}t^{-1}\left(\frac{t+1}{\sqrt{2}}+\sqrt{t}\right)^{-1},$$

$$K_{21}(1,t)=-a_{\neq}^{-1}(-1,-t)b_1^{-1}(-t)=-(4\pi)^{-1}e^{i\frac{\pi}{4}}t^{-1}\left(\frac{1-t}{\sqrt{2}}+i\sqrt{t}\right)^{-1},$$

$$K_{22}(1,t)=a_{\neq}^{-1}(1,-t)b_1^{-1}(-t)=-(4\pi)^{-1}e^{i\frac{\pi}{4}}t^{-1}\left(\frac{t+1}{\sqrt{2}}+\sqrt{t}\right)^{-1}.$$

Analogously we calculate $M_{ij}(t,1)$:

$$M_{11}(t,1)=a_{\neq}^{-1}(t,1)b^{-1}(1)=-(4\pi)^{-1}e^{-i\frac{\pi}{4}}\left(\frac{1-t}{\sqrt{2}}+i\sqrt{t}\right)^{-1}=-itK_{21}(1,t),$$

$$M_{12}(t,1)=a_{\neq}^{-1}(-t,1)b^{-1}(-t)=a_{\neq}^{-1}(-1,t)b^{-1}(-1)=$$

$$= a_{\neq}^{-1}(-1,t)(4\pi)^{-1}e^{i\frac{\pi}{4}} = (4\pi)^{-1}e^{i\frac{\pi}{4}}\left(\frac{t+1}{\sqrt{2}}+\sqrt{t}\right)^{-1} = -tK_{22}(1,t),$$

$$M_{21}(t,1) = a_{\neq}^{-1}(-t,-1)b^{-1}(-t) = a_{\neq}^{-1}(1,t)b^{-1}(-1) =$$

$$= (4\pi)^{-1}e^{i\frac{\pi}{4}}\left(\frac{t-1}{\sqrt{2}}+i\sqrt{t}\right)^{-1} = itK_{11}(1,t),$$

$$M_{22}(t,1) = a_{\neq}^{-1}(t,-1)b^{-1}(t) = a_{\neq}^{-1}(1,-t)b^{-1}(1) =$$

$$= (4\pi)^{-1}e^{-i\frac{\pi}{4}}\left(\frac{t+1}{\sqrt{2}}+\sqrt{t}\right)^{-1} = -tK_{12}(1,t).$$

From the computational point of view it is convenient finding the Mellin transforms of functions $M_{ij}(t,1)$, $i,j = 1,2$, and then taking into account properties of the Mellin transform (see Appendix 3)

$$\widehat{t^{-1}f(t)} = \hat{f}(\lambda-1)$$

to obtain the Mellin transform of functions $K_{ij}(1,t)$.

We will use "prepared" Mellin transforms again (see formulas (10.1.8′) and (10.1.8″)).

Let us write

$$\int_0^{+\infty} \frac{t^{\lambda-1}dt}{\frac{t-1}{\sqrt{2}}+i\sqrt{t}} = 2\sqrt{2}\int_0^{+\infty}\frac{y^{2\lambda-1}dy}{y^2+\sqrt{2}iy-1},$$

and note that the representation holds:

$$\frac{1}{y^2+\sqrt{2}iy-1} = \frac{1}{\sqrt{2}y-1+i} - \frac{1}{\sqrt{2}y+1+i},$$

for which its summands can be conveniently written in form

$$\frac{1}{\sqrt{2}y+i-1} = -\frac{i+1}{2}\left(1+\frac{\sqrt{2}}{i-1}y\right)^{-1},$$

$$\frac{1}{\sqrt{2}y+i+1} = -\frac{i-1}{2}\left(1+\frac{\sqrt{2}}{i+1}y\right)^{-1},$$

and after this acording to formulas (10.1.8′), (10.1.8″) we will obtain

$$\int_0^{+\infty}\frac{t^{\lambda-1}dt}{a_{\neq}(1,t)} = 2\sqrt{2}\left(\frac{i-1}{2}\frac{\pi}{\left(\frac{\sqrt{2}}{i+1}\right)^{2\lambda}}\operatorname{cosec}(2\pi\lambda) - \right.$$

$$\left. -\frac{i+1}{2}\frac{\pi}{\left(\frac{\sqrt{2}}{i-1}\right)^{2\lambda}}\operatorname{cosec}(2\pi\lambda)\right) =$$

$$= 2\pi \operatorname{cosec}(2\pi\lambda)\left(\left(\frac{-1+i}{2}\right)\left(\frac{i+1}{\sqrt{2}}\right)^{2\lambda} - \left(\frac{1+i}{\sqrt{2}}\right)\left(\frac{-1+i}{\sqrt{2}}\right)^{2\lambda}\right) =$$

$$= -2\pi \operatorname{cosec}(2\pi\lambda)\left(e^{i\frac{\pi}{4}(2\lambda-1)} - e^{i\frac{3\pi}{4}(2\lambda-1)}\right),$$

from which follows that

$$\hat{M}_{21}(\lambda) = -\frac{1}{2}e^{i\frac{\pi}{4}} \operatorname{cosec}(2\pi\lambda)\left(e^{i\frac{\pi}{4}(2\lambda-1)} - e^{i\frac{3\pi}{4}(2\lambda-1)}\right),$$

and then

$$\hat{K}_{11}(\lambda) = -i\hat{M}_{21}(\lambda-1),$$

$$\hat{K}_{11}(\lambda) = \frac{i}{2}e^{i\frac{\pi}{4}} \operatorname{cosec}(2\pi(\lambda-1))\left(e^{i\frac{\pi}{4}(2\lambda-3)} - e^{i\frac{3\pi}{4}(2\lambda-3)}\right) =$$

$$= \frac{i}{2}\operatorname{cosec}(2\pi\lambda)e^{i\frac{\pi}{4}}\left(e^{i\frac{\pi}{4}(2\lambda-1)}e^{-i\frac{\pi}{2}} - e^{i\frac{3\pi}{4}(2\lambda-1)}e^{-i\frac{3\pi}{2}}\right) =$$

$$= \frac{i}{2}\operatorname{cosec}(2\pi\lambda)e^{i\frac{\pi}{4}}\left(-ie^{i\frac{\pi}{4}(2\lambda-1)} - ie^{i\frac{3\pi}{4}(2\lambda-1)}\right) =$$

$$= \frac{1}{2}\operatorname{cosec}(2\pi\lambda)e^{i\frac{\pi}{4}}\left(e^{i\frac{\pi}{4}(2\lambda-1)} + e^{i\frac{3\pi}{4}(2\lambda-1)}\right).$$

Further,

$$\int_0^{+\infty} \frac{t^{\lambda-1}dt}{\frac{t+1}{\sqrt{2}} + \sqrt{t}} = \sqrt{2}\int_0^{+\infty} \frac{t^{\lambda-1}dt}{t+\sqrt{2t}+1} =$$

(change $\sqrt{t} = y$)

$$= \sqrt{2}\int_0^{+\infty} \frac{y^{2(\lambda-1)}2ydy}{y^2+\sqrt{2}y+1} = 2\sqrt{2}\int_0^{+\infty} \frac{y^{2\lambda-1}dy}{y^2+\sqrt{2}y+1}.$$

The roots of trinomial

$$y^2+\sqrt{2}y+1$$

are equal to $y_{1;2} = \frac{-\sqrt{2}\pm i\sqrt{2}}{2} = \frac{-1\pm i}{\sqrt{2}}$, and hence

$$y^2+\sqrt{2}y+1 = \left(y+\frac{1+i}{\sqrt{2}}\right)\left(y+\frac{1-i}{\sqrt{2}}\right),$$

$$\frac{1}{y^2+\sqrt{2}y+1} = \left(\frac{1}{y+\frac{1-i}{\sqrt{2}}} - \frac{1}{y+\frac{1+i}{\sqrt{2}}}\right)\left(-\frac{i}{\sqrt{2}}\right).$$

Using formulas (10.1.8′), (10.1.8″) again we obtain

$$\int_0^{+\infty} \frac{t^{\lambda-1}dt}{a_{\neq}(-1,t)} = 2i\left(\frac{\sqrt{2}}{1+i}\frac{\pi}{\left(\frac{\sqrt{2}}{1+i}\right)^{2\lambda}} - \frac{\sqrt{2}}{1-i}\frac{\pi}{\left(\frac{\sqrt{2}}{1-i}\right)^{2\lambda}}\right)\operatorname{cosec}(2\pi\lambda) =$$

$$= 2\pi i \operatorname{cosec}(2\pi\lambda)\left(e^{i\frac{\pi}{4}(2\lambda-1)} - e^{-i\frac{\pi}{4}(2\lambda-1)}\right) =$$

$$= -4\pi \operatorname{cosec}(2\pi\lambda)\sin\frac{\pi}{4}(2\lambda-1).$$

Then

$$\hat{M}_{22}(\lambda) = -e^{-i\frac{\pi}{4}}\operatorname{cosec}(2\pi\lambda)\sin\frac{\pi}{4}(2\lambda-1),$$

and, hence

$$\hat{K}_{12}(\lambda) = -\hat{M}_{22}(\lambda-1) = -e^{-i\frac{\pi}{4}}\operatorname{cosec}(2\pi\lambda)\cos\frac{\pi}{4}(2\lambda-1).$$

The next to the last Mellin transform is

$$\int_0^{+\infty}\frac{t^{\lambda-1}dt}{a_{\neq}(-1,t)} = -4\pi\operatorname{cosec}(2\pi\lambda)\sin\frac{\pi}{4}(2\lambda-1),$$

and then

$$\hat{M}_{12}(\lambda) = -e^{i\frac{\pi}{4}}\operatorname{cosec}(2\pi\lambda)\sin\frac{\pi}{4}(2\lambda-1),$$

$$\hat{K}_{22}(\lambda) = -\hat{M}_{12}(\lambda-1) = -e^{i\frac{\pi}{4}}\operatorname{cosec}(2\pi\lambda)\cos\frac{\pi}{4}(2\lambda-1).$$

The last Mellin transform:

$$\int_0^{+\infty}\frac{t^{\lambda-1}dt}{a_{\neq}(-1,-t)} = \int_0^{+\infty}\frac{t^{\lambda-1}dt}{\frac{1-t}{\sqrt{2}}+i\sqrt{t}} = -\sqrt{2}\int_0^{+\infty}\frac{t^{\lambda-1}dt}{t-\sqrt{2}i\sqrt{t}-1} =$$

$$= -2\sqrt{2}\int_0^{+\infty}\frac{y^{2\lambda-1}dy}{y^2-\sqrt{2}iy-1}.$$

Roots of trinomial

$$y^2-\sqrt{2}iy-1$$

are $y_{1;2} = \frac{\pm\sqrt{2}+\sqrt{2}i}{2} = \frac{\pm 1+i}{\sqrt{2}}$, and

$$y^2-\sqrt{2}iy-1 = \left(y-\frac{1+i}{\sqrt{2}}\right)\left(y-\frac{i-1}{\sqrt{2}}\right),$$

$$\frac{1}{y^2-\sqrt{2}iy-1} = -\frac{1}{\sqrt{2}}\left(\frac{1}{y-\frac{i-1}{\sqrt{2}}} - \frac{1}{y-\frac{1+i}{\sqrt{2}}}\right) =$$

$$= \left(\frac{\left(\frac{\sqrt{2}}{-1-i}\right)}{\frac{\sqrt{2}}{(-1-i)}y+1} - \frac{\left(\frac{\sqrt{2}}{1-i}\right)}{\frac{\sqrt{2}}{(1-i)}y+1}\right)\frac{1}{\sqrt{2}},$$

and applying again the pair of formulas (10.1.8′), (10.1.8″), we obtain

$$\int_0^{+\infty} \frac{t^{\lambda-1}dt}{a_{\neq}(-1,-t)} = -2\sqrt{2}\left[\frac{1}{-1-i}\frac{\pi}{\left(\frac{\sqrt{2}}{-1-i}\right)^{2\lambda}}\operatorname{cosec}(2\pi\lambda) - \right.$$

$$\left. -\frac{1}{1-i}\frac{\pi}{\left(\frac{\sqrt{2}}{1-i}\right)^{2\lambda}}\operatorname{cosec}(2\pi\lambda)\right] =$$

$$= -2\pi\operatorname{cosec}(2\pi\lambda)\left(e^{-i\frac{3\pi}{4}(2\lambda-1)} - e^{-i\frac{\pi}{4}(2\lambda-1)}\right).$$

So,

$$\hat{M}_{11}(\lambda) = \frac{1}{2}e^{-i\frac{\pi}{4}}\operatorname{cosec}(2\pi\lambda)\left(e^{-i\frac{3\pi}{4}(2\lambda-1)} - e^{-i\frac{\pi}{4}(2\lambda-1)}\right),$$

and then

$$\hat{K}_{21}(\lambda) = i\hat{M}_{11}(\lambda-1) = \frac{1}{2}ie^{-i\frac{\pi}{4}}\operatorname{cosec}(2\pi\lambda)\left(e^{-i\frac{3\pi}{4}(2\lambda-1)} - e^{-i\frac{\pi}{4}(2\lambda-1)}\right) =$$

$$= \frac{1}{2}e^{i\frac{\pi}{4}}\operatorname{cosec}(2\pi\lambda)\left(-ie^{-i\frac{3\pi}{4}(2\lambda-1)} - ie^{-i\frac{\pi}{4}(2\lambda-1)}\right) =$$

$$= -\frac{1}{2}ie^{i\frac{\pi}{4}}\operatorname{cosec}(2\pi\lambda)\left(e^{-i\frac{3\pi}{4}(2\lambda-1)} + e^{-i\frac{\pi}{4}(2\lambda-1)}\right) =$$

$$= \frac{1}{2}e^{-i\frac{\pi}{4}}\operatorname{cosec}(2\pi\lambda)\left(e^{-i\frac{3\pi}{4}(2\lambda-1)} + e^{-i\frac{\pi}{4}(2\lambda-1)}\right).$$

By Lemma 10.1.1

$$\Delta_A(\lambda) = \det K(\lambda)\cdot\det M(\lambda) - \operatorname{sp}[K(\lambda)\cdot M(\lambda)] + 1,$$

where

$$K(\lambda) = \begin{pmatrix} \hat{K}_{11}(\lambda) & \hat{K}_{12}(\lambda) \\ \hat{K}_{22}(\lambda) & \hat{K}_{21}(\lambda) \end{pmatrix}, \quad M(\lambda) = \begin{pmatrix} \hat{M}_{11}(\lambda) & \hat{M}_{22}(\lambda) \\ \hat{M}_{12}(\lambda) & \hat{M}_{21}(\lambda) \end{pmatrix}.$$

Our calculations:

$$\det K(\lambda) = \hat{K}_{11}(\lambda)\hat{K}_{21}(\lambda) - \hat{K}_{12}(\lambda)\hat{K}_{22}(\lambda) =$$

$$= \frac{1}{4}\operatorname{cosec}^2(2\pi\lambda)\left(e^{i\frac{\pi}{4}(2\lambda-1)} + e^{i\frac{3\pi}{4}(2\lambda-1)}\right)\left(e^{-i\frac{3\pi}{4}(2\lambda-1)} + e^{-i\frac{\pi}{4}(2\lambda-1)}\right) -$$

$$-\operatorname{cosec}^2(2\pi\lambda)\cos^2\frac{\pi}{4}(2\lambda-1) =$$

$$= \frac{1}{4}\operatorname{cosec}^2(2\pi\lambda)\left(e^{-i\frac{\pi}{2}(2\lambda-1)} + 1 + 1 + e^{i\frac{\pi}{2}(2\lambda-1)}\right) -$$

$$-\operatorname{cosec}^2(2\pi\lambda)\cos^2\frac{\pi}{4}(2\lambda-1) = \frac{1}{2}\operatorname{cosec}^2(2\pi\lambda)\left(1+\cos\frac{\pi}{2}(2\lambda-1)\right) -$$

$$-\operatorname{cosec}^2(2\pi\lambda)\cos^2\frac{\pi}{4}(2\lambda-1) = \operatorname{cosec}^2(2\pi\lambda)\cos^2\frac{\pi}{4}(2\lambda-1)-$$

$$-\operatorname{cosec}^2(2\pi\lambda)\cos^2\frac{\pi}{4}(2\lambda-1) = 0;$$

$$\det M(\lambda) = \hat{M}_{11}(\lambda)\hat{M}_{21}(\lambda) - \hat{M}_{12}(\lambda)\hat{M}_{22}(\lambda) =$$

$$= -\frac{1}{4}\operatorname{cosec}^2(2\pi\lambda)\left(e^{-i\frac{3\pi}{4}(2\lambda-1)} - e^{-i\frac{\pi}{4}(2\lambda-1)}\right)\left(e^{i\frac{\pi}{4}(2\lambda-1)} - e^{i\frac{3\pi}{4}(2\lambda-1)}\right) -$$

$$-\operatorname{cosec}^2(2\pi\lambda)\sin^2\frac{\pi}{4}(2\lambda-1) = .$$

$$= -\frac{1}{4}\operatorname{cosec}^2(2\pi\lambda)\left(e^{-i\frac{\pi}{2}(2\lambda-1)} - 1 - 1 + e^{i\frac{\pi}{2}(2\lambda-1)}\right) -$$

$$-\operatorname{cosec}^2(2\pi\lambda)\sin^2\frac{\pi}{4}(2\lambda-1) =$$

$$= -\frac{1}{2}\operatorname{cosec}^2(2\pi\lambda)\left(\cos\frac{\pi}{2}(2\lambda-1) - 1\right) - \operatorname{cosec}^2(2\pi\lambda)\sin^2\frac{\pi}{4}(2\lambda-1) =$$

$$= \operatorname{cosec}^2(2\pi\lambda)\sin^2\frac{\pi}{4}(2\lambda-1) - \operatorname{cosec}^2(2\pi\lambda)\sin^2\frac{\pi}{4}(2\lambda-1) = 0.$$

Let's calculate the trace of matrix $K(\lambda)\cdot M(\lambda)$:

$$\operatorname{sp}[K(\lambda)\cdot M(\lambda)] = \hat{K}_{11}(\lambda)\hat{M}_{11}(\lambda) + \hat{K}_{12}(\lambda)\hat{M}_{12}(\lambda)+$$

$$+\hat{K}_{22}(\lambda)\hat{M}_{22}(\lambda) + \hat{K}_{21}(\lambda)\hat{M}_{21}(\lambda) =$$

$$= \frac{1}{4}\operatorname{cosec}^2(2\pi\lambda)\left(e^{i\frac{\pi}{4}(2\lambda-1)} + e^{i\frac{3\pi}{4}(2\lambda-1)}\right)\left(e^{-i\frac{3\pi}{4}(2\lambda-1)} - e^{-i\frac{\pi}{4}(2\lambda-1)}\right) +$$

$$+\operatorname{cosec}^2(2\pi\lambda)\sin\frac{\pi}{4}(2\lambda-1)\cos\frac{\pi}{4}(2\lambda-1)+$$

$$+\operatorname{cosec}^2(2\pi\lambda)\sin\frac{\pi}{4}(2\lambda-1)\cos\frac{\pi}{4}(2\lambda-1)-$$

$$-\frac{1}{4}\operatorname{cosec}^2(2\pi\lambda)\left(e^{-i\frac{3\pi}{4}(2\lambda-1)} + e^{-i\frac{\pi}{4}(2\lambda-1)}\right)\left(e^{i\frac{\pi}{4}(2\lambda-1)} - e^{i\frac{3\pi}{4}(2\lambda-1)}\right) =$$

$$= \frac{1}{4}\operatorname{cosec}^2(2\pi\lambda)\left(e^{-i\frac{\pi}{2}(2\lambda-1)} + 1 - 1 - e^{i\frac{\pi}{2}(2\lambda-1)}\right) -$$

$$-\frac{1}{4}\operatorname{cosec}^2(2\pi\lambda)\left(e^{-i\frac{\pi}{2}(2\lambda-1)} + 1 - 1 - e^{i\frac{\pi}{2}(2\lambda-1)}\right) + \operatorname{cosec}^2(2\pi\lambda)\sin\frac{\pi}{2}(2\lambda-1),$$

from which by Lemma 10.1.1 follows that

$$\Delta_A(\lambda) = 1 - \operatorname{cosec}^2(2\pi\lambda)\sin\frac{\pi}{2}(2\lambda-1). \tag{10.2.8}$$

As before we will denote $[v]_{s,\alpha}$ the norm of function v in space $H_{s,\alpha}(\mathbb{R})$:

$$[v]_{s,\alpha}^2 = \int_{-\infty}^{+\infty} |\tilde{v}(t)|^2\,|t|^{2\alpha}(1+|t|)^{2(s-\alpha)}dt,$$

$$H_{s,\alpha}(\mathbb{R}_+) \equiv \{u \in H_{s,\alpha}(\mathbb{R}) : \ \operatorname{supp} u \subset [0; +\infty)\}.$$

Theorem 10.2.1. *Let* $g_1 \in H_{s-3/2,-1}(\mathbb{R}_+)$, $g_2 \in H_{s-1/2}(\mathbb{R}_+)$, $s > 3/2$. *The problem (10.2.1), (10.2.2) has a unique solution in space* $H_{s,1}(C^1_+)$ *and a priori estimate holds*

$$||u_+||_{s,1} \leq c\left([g_1]_{s-3/2,-1} + [g_2]_{s-1/2}\right).$$

Proof. The 4×4 - system of linear algebraic equations will have a unique solution, and together with it the 4×4 - system of linear integral equations on half-axis (and hence, the 2×2 - system (10.2.7) of linear integral equations on the whole axis too) will have a unique solution if the $\Delta_A(\lambda) \neq 0$, $\operatorname{Re}\lambda = 1/2$.

Let us consider the determinant $\Delta_A(\lambda)$ (10.2.8) explicitly. It is not difficult to see that $\Delta_A(\lambda)$ is a analytic function in the whole complex plane $\mathbb{C}$ excepting the real points $\lambda = 0;\ \pm 1/2;\ \pm 3/2; \ldots$, in which $\Delta_A(\lambda)$ has poles (of second order at zero and first order in other points), and these points will be zeroes of function $\Delta_A^{-1}(\lambda)$, which is analytic in $\mathbb{C}$ excepting points in which $\Delta_A(\lambda)$ is equal to zero. These points are roots of equation

$$1 - \operatorname{cosec}^2(2\pi\lambda)\sin\frac{\pi}{2}(2\lambda - 1) = 0. \tag{10.2.9}$$

By non-complicated transformations one reduces the equation (10.2.9) to a cubic equation with respect to $\cos\pi\lambda \equiv z$ of type

$$4z^3 + 4z + 1 = 0,$$

from which follows that equation (10.2.9) has sets of roots corresponding to roots z_1, z_2, z_3 of the last equation, namely $\frac{1}{\pi}\operatorname{Arccos} z_1$, $\frac{1}{\pi}\operatorname{Arccos} z_2$, $\frac{1}{\pi}\operatorname{Arccos} z_3$, which are situated on straight lines parallel to the real axis

$$\begin{aligned} \operatorname{Arccos} z_j &= \pm i\ln t_j + 2\pi k, \quad k = 0; \pm 1; \pm 2; \ldots \\ t_j &= \sqrt{z_j^2 - 1}, \quad j = 1, 2, 3, \text{ and} \\ \operatorname{Re}\operatorname{Arccos} z_j &= \pm\arg t_j + 2\pi k, \quad k = 0; \pm 1; \pm 2; \ldots . \end{aligned}$$

So, the set T of roots of equation (10.2.9) consists of complex numbers with real part of type

$$\operatorname{Re}\lambda_j^{(k)} = \pm\frac{1}{\pi}\arg t_j + 2k, \quad k = 0; \pm 1; \pm 2; \ldots, \quad j = 1, 2, 3,$$

and the function $\Delta_A^{-1}(\lambda)$ is analytic in $\mathbb{C} \setminus T$, where T is set of poles of function $\Delta_A^{-1}(\lambda)$.

On the straight line $\operatorname{Re}\lambda = 1/2$ the determinant $\Delta_A(\lambda)$ has no zeroes, and $\Delta_A(\lambda) \to 1$ under $\operatorname{Im}\lambda \to \infty$. In fact,

$$\Delta_A(1/2 + ib) = 1 - \frac{\sin(\pi ib)}{\sin^2(2\pi ib)} = 1 - \frac{1}{4\sin(\pi ib)\cos^2(\pi ib)},$$

and $\Delta_A(1/2 + ib)$ can not take the vanishing value, because $\sin(\pi ib)$ is a complex number, but $\cos(\pi ib)$ is a real number. Hence, $\Delta_A^{-1}(1/2+ib)$ is a bounded function.

Using the result from Section 9.3 (Lemma 9.3.2) we have

$$||u_+||_{s,1} = ||U_+||_{s,1} = ||\tilde{U}_+||_{s,1} \leq$$

$$\leq c_1 \left([c]_{s-1/2} + [d]_{s-1/2} \right),$$

and it is left to obtain estimates for $[c]_{s-1/2}$, $[d]_{s-1/2}$ via g_1, g_2 respectively.

Let us introduce the following notations. The matrix of 4×4 - system of linear algebraic equations with determinant $\Delta_A(\lambda)$ we will denote by $A_4(\lambda)$, the vector with components $\hat{c}_0(\lambda)$, $\hat{c}_1(\lambda)$, $\hat{d}_0(\lambda)$, $\hat{d}_1(\lambda)$ will be $\hat{C}(\lambda)$, and the vector with components $\hat{r}_{10}(\lambda)$, $\hat{r}_{11}(\lambda)$, $\hat{r}_{20}(\lambda)$, $\hat{r}_{21}(\lambda)$ will be $\hat{V}(\lambda)$. Then

$$\hat{C}(\lambda) = \hat{A}_4^{-1}(\lambda)\hat{V}(\lambda),$$

and, since the components of matrix $\hat{A}_4^{-1}(\lambda)$ are bounded on the straight line $\operatorname{Re}\lambda = 1/2$ then

$$||\hat{C}(\lambda)||_0 \leq c'||\hat{V}(\lambda)||_0,$$

where

$$||\hat{C}(\lambda)||_0 = \frac{1}{2\pi i} \int\limits_{1/2-i\infty}^{1/2+i\infty} |\hat{C}(\lambda)|^2 d\lambda,$$

and from which by Parceval equality

$$\int\limits_0^{+\infty} |C(x)|^2 dx = \int\limits_{1/2-i\infty}^{1/2+i\infty} |\hat{C}(\lambda)|^2 d\lambda$$

we obtain

$$\int\limits_0^{+\infty} |C(x)|^2 dx \leq c' \int\limits_0^{+\infty} |V(x)|^2 dx.$$

Returning to $c(x)$, $d(x)$ we have

$$[c]_0 \leq c_1 \left(\left[b_1^{-1} g_1\right]_0 + \left[b^{-1} g_2\right]_0 \right),$$

$$[d]_0 \leq c_2 \left(\left[b_1^{-1} g_1\right]_0 + \left[b^{-1} g_2\right]_0 \right),$$

and this implies by virtue of boundedness for b^{-1}

$$\left[b^{-1} g_2\right]_0 \leq c_1' \, [g_2]_0 \leq c_1' \, [g_2]_{s-1/2} \quad (s - 1/2 > 0).$$

The second estimate:

$$\left[b_1^{-1} g_1\right]_0^2 \leq c_1' \int\limits_{-\infty}^{+\infty} \frac{|\tilde{g}_1(t)|^2 dt}{|t|^2} = c_1' \int\limits_{-\infty}^{+\infty} \left(\frac{|t|}{1 + |t|} \right)^{-2} |\tilde{g}_1(t)|^2 \frac{dt}{(1 + |t|)^2} \leq$$

$$\leq c_1' \int_{-\infty}^{+\infty} \left(\frac{|t|}{1+|t|}\right)^{-2} |\tilde{g}_1(t)|^2(1+|t|)^{2(s-3/2)}dt = c_1' \, [g_1]^2_{s-3/2,-1} \, .$$

Let's consider $[c]_{s-1/2}$.

$$[c]^2_{s-1/2} = \int_{-\infty}^{+\infty} |\tilde{c}(t)|^2(1+|t|)^{2(s-1/2)}dt =$$

$$= \left(\int_{|t|<1} + \int_{|t|\geq 1} \right) |\tilde{c}(t)|^2(1+|t|)^{2(s-1/2)}dt.$$

If $|t| < 1$ then $1 \leq 1+|t| \leq 2$, i.e. $1 \sim 1+|t|$, and in case $|t| \geq 1$ we have $|t| \leq 1+|t| \leq 2|t|$, i.e. $|t| \sim 1+|t|$, hence

$$[c]^2_{s-1/2} \leq \text{const} \left([c]^2_0 + \int_{-\infty}^{+\infty} |\tilde{c}(t)|^2(1+|t|)^{2(s-1/2)}dt \right).$$

We denote

$$\left(\int_{-\infty}^{+\infty} |\tilde{c}(t)|^2(1+|t|)^{2\alpha}dt \right)^{1/2} \equiv \langle c \rangle_\alpha,$$

then

$$[c]^2_{s-1/2} \leq \text{const} \left([c]^2_0 + \langle c \rangle^2_{s-1/2} \right).$$

Let's assume s is such that the straight line $\operatorname{Re}\lambda = s$ doesn't contain zeroes of determinant $\Delta_A(\lambda)$. Then $\Delta_A^{-1}(\lambda)$ is bounded on this straight line, and

$$\langle \hat{C}(\lambda) \rangle_{s-1/2} = \langle \hat{A}_4^{-1}(\lambda)\hat{V}(\lambda) \rangle_{s-1/2} \leq c' \langle \hat{V}(\lambda) \rangle_{s-1/2},$$

where

$$\langle \hat{C}(\lambda) \rangle^2_{s-1/2} = \frac{1}{2\pi i} \int_{s-i\infty}^{s+i\infty} |\hat{C}(\lambda)|^2 d\lambda,$$

and by virtue of Parceval equality

$$\int_0^{+\infty} |x|^{2s-1}|C(x)|^2dx = \frac{1}{2\pi i} \int_{s-i\infty}^{s+i\infty} |\hat{C}(\lambda)|^2 d\lambda$$

we obtain

$$\int_0^{+\infty} |x|^{2s-1}|C(x)|^2dx \leq c' \int_0^{+\infty} |x|^{2s-1}|V(x)|^2dx,$$

and further,

$$\langle c\rangle^2_{s-1/2} \le c_1'\left(\left[b_1^{-1}g_1\right]^2_{s-1/2} + \left[b^{-1}g_2\right]^2_{s-1/2}\right) \le$$

$$\le c_1'\left([g_2]^2_{s-1/2} + \int\limits_{-\infty}^{+\infty} \frac{|\tilde{g}_1(t)|^2(1+|t|)^{2(s-1/2)}dt}{|t|^2}\right) =$$

$$= c_1'\left([g_2]^2_{s-1/2} + \int\limits_{-\infty}^{+\infty} \left(\frac{|t|}{1+|t|}\right)^{-2} |\tilde{g}_1(t)|^2(1+|t|)^{2(s-3/2)}dt = \right.$$

$$\left. = c_1'\left([g_1]^2_{s-3/2,-1} + [g_2]^2_{s-1/2}\right)\right. .$$

Collecting the estimates above we obtain the statement of the Theorem 10.2.1.

If the straight line $\operatorname{Re}\lambda = s$ contains zeroes of determinant $\Delta_A(\lambda)$ then by virtue of their structure (see above) the straight line $\operatorname{Re}\lambda = s - \varepsilon$ does not have zeroes of $\Delta_A(\lambda)$ for small enough $\varepsilon > 0$. Using an analogous argument with $s - \varepsilon$ instead of s and taking into account that ($s > 3/2$)

$$||u||_{s-\varepsilon} \le ||u||_s$$

we obtain the estimate needed. ∎

If one makes bold to consider the Cauchy problem for equation (10.2.1), (i.e., such case when the Dirichlet and Neumann conditions are given on just the same angle side ∂C_+^1), it is easy to see that determinant $\Delta_A(\lambda)$ will be like such

$$\Delta_A(\lambda) = \begin{vmatrix} \hat{K}_{11}(\lambda) & \hat{K}_{12}(\lambda) & 1 & 0 \\ \hat{K}_{22}(\lambda) & \hat{K}_{21}(\lambda) & 0 & 1 \\ \hat{M}_{11}(\lambda) & \hat{M}_{22}(\lambda) & 1 & 0 \\ \hat{M}_{12}(\lambda) & \hat{M}_{21}(\lambda) & 0 & 1 \end{vmatrix}.$$

The explicit form for function $\hat{K}_{ij}(\lambda)$ (and their relation to $\hat{M}_{ij}(\lambda)$) permits us to conclude quickly that under $\operatorname{Re}\lambda = \text{const}$, $\operatorname{Im}\lambda \to \pm\infty$ we have $\hat{K}_{ij}(\lambda) \to 0$ from which follows that $\Delta_A(\lambda) \to 0$.

So, the corresponding 4×4 - system of linear algebraic equations (under arbitrary right sides) is unsolvable, and together with it the Cauchy problem is unsolvable too.

11. Problems with potentials

Here we consider again the equation

$$(Au_+)(x) = f(x), \quad x \in C_+^a, \tag{11.1}$$

to show how one can use Theorem 8.1.3 and what kind of boundary value problems correspond to this case. We assume $æ - s = -1 + \delta$, $|\delta| < 1/2$, æ is index of wave factorization of symbol $A(\xi) \in C_\alpha^\infty$.

As it is shown in Section 8.1 the solution $u_+ \in H_s(C_+^a)$ exists ($f \in H'_{s-\alpha}(C_+^a)$) and it can be represented in the form (after a change which transforms C_+^1 into first quadrant K)

$$\tilde{W}(x) = \tilde{U}(x) + \tilde{V}_1(x_2)B_1(x) + \tilde{V}_2(x_1)B_2(x), \tag{11.2}$$

where $B_1(x)$, $B_2(x)$ are explicitly expressed via symbol $A_{\neq}(\xi)$, $U(x) \in H_s(K)$, $V_1(x_2)$, $V_2(x_1)$ are the functions belonging to class $H_{1/2-\delta}(\mathbb{R}_+)$, and so the last summands of formula (11.2) are some potentials. Let us write out the expression for B_1, B_2. By formula (8.1.12) we have

$$B_j(x) = (x_j - x_{3-j} - i)^{-1} a_{\neq}(x), \quad j = 1, 2.$$

Finally, $W(x)$ is defined by formula

$$\tilde{w} = A_{\neq}^{-1} G'_2 A_{=}^{-1} \tilde{\ell f}. \tag{11.3}$$

(All notations are taken from n° 8.1).

By methods which are analogous to those considered in [92] it may be proved that representation of solution $W(x)$ in form (11.2) is unique. It permits us to consider more the general problem with potential type operators in form

$$(Au)(x) + B_1\left(v_1(x_2) \otimes \delta(x_1)\right) + \left(v_2(x_1) \otimes \delta(x_2)\right) = f(x) \tag{11.4}$$

in quadrant K on plane, where $\otimes$ denotes direct product of distributions.

The formal scheme for solving of equation (11.4) follows. Transferring the last two summands to the right-hand side, meaning that they are known, we obtain the equation of type (11.1) in a quadrant for which its solvability theory is described in Section 8.1, Theorem 8.1.3: in order that this equation have a solution belonging to $H_s(K)$ it is necessary and sufficient that the conditions

$$\left.(A_{=}^{-1} g)(x)\right|_{\substack{x_1 = 0 \\ x_2 > 0}} = 0, \quad \left.(A_{=}^{-1} g)(x)\right|_{\substack{x_1 > 0 \\ x_2 = 0}} = 0 \tag{11.5}$$

are fulfilled, where $g(x) = (\ell f)(x) - B_1\left(v_1(x_2) \otimes \delta(x_1)\right) - B_2\left(v_2(x_1) \otimes \delta(x_2)\right)$.

These conditions written in Fourier images have the following form (see (3.1.2)):

$$\int_{-\infty}^{+\infty} A_{=}^{-1}(\xi)\left[\tilde{\ell f}(\xi) - B_1(\xi)\tilde{v}_1(\xi_2) - B_2(\xi)\tilde{v}_2(\xi_1)\right] d\xi_j = 0, \tag{11.6}$$

$$j = 1, 2.$$

Let us denote

$$\int_{-\infty}^{+\infty} \frac{\tilde{\ell f}(\xi)}{A_{=}(\xi)} d\xi_j = g_j(\xi_{3-j}), \quad j = 1, 2,$$

$$B_i(\xi)A_{=}^{-1}(\xi) = K_i'(\xi), \quad i = 1, 2.$$

As a result the equality (11.6) gives a system of two linear integral equations for finding of $\tilde{v}_1(\xi)$, $\tilde{v}_2(\xi)$:

$$\int_{-\infty}^{+\infty} K_1'(\xi_1, \xi_2)\tilde{v}_1(\xi_2)d\xi_j + \int_{-\infty}^{+\infty} K_2'(\xi_1, \xi_2)\tilde{v}_2(\xi_2)d\xi_j = g_j(\xi_{3-j}), \quad j = 1, 2.$$

Let us denote also

$$\int_{-\infty}^{+\infty} K_j'(\xi_1, \xi_2)d\xi_j = b_j(\xi_{3-j}), \quad j = 1, 2,$$

and assume that

$$b_1(\xi_2) \neq 0, \ \ \forall \xi_2 \neq 0; \quad b_2(\xi_1) \neq 0, \ \ \forall \xi_1 \neq 0. \tag{11.7}$$

Then the last system can be written in form ($K_2(\xi_1, \xi_2) = K_2'(\xi_1, \xi_2)b_1^{-1}(\xi_2)$, $K_1(\xi_1, \xi_2) = K_1'(\xi_1, \xi_2)b_2^{-1}(\xi_1)$, $g_1'(\xi_2) = b_1^{-1}(\xi_2)g_1(\xi_2)$, $g_2'(\xi_1) = b_2^{-1}(\xi_1)g_2(\xi_1)$)

$$\begin{cases} \tilde{v}_1(\xi_2) + \displaystyle\int_{-\infty}^{+\infty} K_2(\xi_1, \xi_2)\tilde{v}_2(\xi_1)d\xi_1 = g_1'(\xi_2) \\ \displaystyle\int_{-\infty}^{+\infty} K_1(\xi_1, \xi_2)\tilde{v}_1(\xi_2)d\xi_2 + \tilde{v}_2(\xi_1) = g_2'(\xi_1) \end{cases} \tag{11.8}$$

So, the equation (11.4) will have unique solution (u, v_1, v_2) if and only if the system (11.8) will be uniquely solvable (and u will be defined by a formula of type (11.3)).

Let the symbol $A(\xi)$ be homogeneous of order α, and symbols $B_1(\xi)$, $B_2(\xi)$ be homogeneous of order β_1, β_2 respectively.

Lemma 11.1. *The kernels $K_i(\xi_1, \xi_2)$, $i = 1, 2$, are homogeneous of order -1 with respect to variables ξ_1, ξ_2.*

Proof. An elmentary verification: $A_{=}^{-1}(\xi)$ is homogeneous of order æ $-\ \alpha$, hence $K_1'(\xi)$ is homogeneous of order $\beta_1 +$ æ $- \alpha$. Then

$$b_1(\lambda\xi_2) = \int_{-\infty}^{+\infty} K_1'(\xi_1, \lambda\xi_2)d\xi_1 = \int_{-\infty}^{+\infty} K_1'(\lambda\xi_1, \lambda\xi_2)\lambda d\xi_1 =$$

$$= \lambda^{\beta_1 + æ - \alpha + 1} b_1(\xi_2),$$

and consequently $K_1'(\xi)b_1^{-1}(\xi_2)$ is homogeneous of order -1. ■

The lemma proved permits use of the Mellin transform for studying solvability of the system (11.8).

As before we introduce the following notations. Let $\tilde{v}_{10}(\xi_2)$ denotes restriction $\tilde{v}_1(\xi_2)$ on $(0;+\infty)$, $\tilde{v}_{11}(\xi_2)$ is restriction $\tilde{v}_1(\xi_2)$ on $(-\infty;0)$. Analogously we define $\tilde{v}_{20}(\xi_1)$, $\tilde{v}_{21}(\xi_1)$, $g_{10}(\xi_2)$, $g_{11}(\xi_2)$, $g_{20}(\xi_1)$, $g_{21}(\xi_1)$. Further, for all $\xi_1, \xi_2 > 0$ we put

$$K_{i11}(\xi_1,\xi_2) = K_i(\xi_1,\xi_2), \quad K_{i12}(\xi_1,\xi_2) = K_i(-\xi_1,\xi_2),$$

$$K_{i22}(\xi_1,\xi_2) = K_i(-\xi_1,-\xi_2), \quad K_{i21}(\xi_1,\xi_2) = K_i(\xi_1,-\xi_2), \quad i = 1,2.$$

Let's consider the second equation of system (11.8),

$$\int_{-\infty}^{0} K_1(\xi_1,\xi_2)\tilde{v}_1(\xi_2)d\xi_2 + \int_{0}^{+\infty} K_1(\xi_1,\xi_2)\tilde{v}_1(\xi_2)d\xi_2 + \tilde{v}_2(\xi_1) = g_2'(\xi_1),$$

and in the first integral change the variable $\xi_2 \mapsto -\xi_2$ obtaining with it

$$\int_{0}^{+\infty} K_1(\xi_1,\xi_2)\tilde{v}_1(\xi_2)d\xi_2 + \int_{0}^{+\infty} K_1(\xi_1,-\xi_2)\tilde{v}_1(-\xi_2)d\xi_2 + \tilde{v}_2(\xi_1) = g_2'(\xi_1),$$

or if $\xi_1 > 0$

$$\int_{0}^{+\infty} K_{111}(\xi_1,\xi_2)\tilde{v}_{10}(\xi_2)d\xi_2 + \int_{0}^{+\infty} K_{121}(\xi_1,\xi_2)\tilde{v}_{11}(\xi_2)d\xi_2 + \tilde{v}_{20}(\xi_1) = g_{20}'(\xi_1), \tag{11.9}$$

but if $\xi_1 < 0$

$$\int_{0}^{+\infty} K_{112}(\xi_1,\xi_2)\tilde{v}_{10}(\xi_2)d\xi_2 + \int_{0}^{+\infty} K_{122}(\xi_1,\xi_2)\tilde{v}_{11}(\xi_2)d\xi_2 + \tilde{v}_{21}(\xi_1) = g_{21}'(\xi_1). \tag{11.10}$$

The first equation of system (11.8):

$$\tilde{v}_1(\xi_2) + \int_{-\infty}^{0} K_2(\xi_1,\xi_2)\tilde{v}_2(\xi_1)d\xi_1 + \int_{0}^{+\infty} K_2(\xi_1,\xi_2)\tilde{v}_2(\xi_1)d\xi_1 = g_1'(\xi_2),$$

$$\tilde{v}_1(\xi_2) + \int_{0}^{+\infty} K_2(\xi_1,\xi_2)\tilde{v}_2(\xi_1)d\xi_1 + \int_{0}^{+\infty} K_2(-\xi_1,\xi_2)\tilde{v}_2(-\xi_2)d\xi_1 = g_1'(\xi_2).$$

Under $\xi_2 > 0$ we have in new notations:

$$\tilde{v}_{10}(\xi_2) + \int_{0}^{+\infty} K_{211}(\xi_1,\xi_2)\tilde{v}_{20}(\xi_1)d\xi_1 + \int_{0}^{+\infty} K_{212}(\xi_1,\xi_2)\tilde{v}_{21}(\xi_1)d\xi_1 = g_{10}'(\xi_2), \tag{11.11}$$

and under $\xi_2 < 0$

$$\tilde{v}_{11}(\xi_2) + \int_{0}^{+\infty} K_{221}(\xi_1,\xi_2)\tilde{v}_{20}(\xi_1)d\xi_1 + \int_{0}^{+\infty} K_{222}(\xi_1,\xi_2)\tilde{v}_{21}(\xi_1)d\xi_1 = g_{11}'(\xi_2). \tag{11.12}$$

So, instead of a 2×2 - system (11.8) of linear integral equations on the whole axis, we have obtained a 4×4 - system of linear integral equations on the half-axis $(0; +\infty)$ with respect to unknown functions $\tilde{v}_{10}$, $\tilde{v}_{11}$, $\tilde{v}_{20}$, $\tilde{v}_{21}$ and this system consists of equations (11.9) – (11.12). Now we will apply the Mellin transform to the last system, and after that by virtue of Lemma 11.1 and properties of the Mellin transform (see Appendix 3) we will obtain a 4×4 - system of linear algebraic equations with matrix

$$K(\lambda) = \begin{pmatrix} \hat{K}_{111}(\lambda) & \hat{K}_{121}(\lambda) & 1 & 0 \\ \hat{K}_{112}(\lambda) & \hat{K}_{122}(\lambda) & 0 & 1 \\ 1 & 0 & \hat{K}_{211}(\lambda) & \hat{K}_{212}(\lambda) \\ 0 & 1 & \hat{K}_{221}(\lambda) & \hat{K}_{222}(\lambda) \end{pmatrix}$$

where for $\hat{K}_{1ij}(\lambda)$ one means the Mellin transform of function $K_{1ij}(t,1)$, and under $\hat{K}_{2ij}(\lambda)$ the Mellin transform of function $K_{2ij}(1,t)$, $t^{-1/2}K_{1ij}$, $t^{-1/2}K_{2ij} \in L(0;+\infty)$, $i,j = 1,2$.

So, for unique determination of $\tilde{v}_{10}$, $\tilde{v}_{11}$, $\tilde{v}_{20}$, $\tilde{v}_{21}$ it is necessary that the determinant of matrix $K(\lambda)$ will be non-vanishing.

We pass to explicit formulations.

We denote by $\tilde{C}_\alpha$ the class of symbols $A(\xi)$ which are homogeneous of order α and satisfy inequality $|A(\xi)| \sim |\xi|^\alpha$.

Lemma 11.2. *Let A be a pseudodifferential operator with symbol $A(\xi) \in \tilde{C}_\alpha$. Then the estimate holds*

$$||Au||_{s-\alpha,\gamma-\alpha} \leq c||u||_{s,\gamma}, \quad \forall u \in S(\mathbb{R}^2),$$

and, hence A extends in continuity to a bounded operator acting from space $H_{s,\gamma}(\mathbb{R}^2)$ into space $H_{s-\alpha,\gamma-\alpha}(\mathbb{R}^2)$.

Proof. Actually,

$$||Au||^2_{s-\alpha,\gamma-\alpha} = \int_{\mathbb{R}^2} |A(\xi)|^2|\xi|^{2(\gamma-\alpha)}(1+|\xi|)^{2(s-\alpha-(\gamma-\alpha))}|\tilde{u}(\xi)|^2 d\xi \leq$$

$$\leq \int_{\mathbb{R}^2} |\xi|^{2\gamma}(1+|\xi|)^{2(s-\gamma)}|\tilde{u}(\xi)|^2 d\xi = c||u||^2_{s,\gamma},$$

from which follows the statement on boundedness of operator $A: \ H_{s,\gamma}(\mathbb{R}^2) \to H_{s-\alpha,\gamma-\alpha}(\mathbb{R}^2)$. ■

Corollary 11.1. *Let A be a pseudodifferential operator with symbol $A(\xi) \in \tilde{C}_\alpha$. Then*

$$||Au||_{s-\alpha} \leq c||u||_{s,\alpha}, \quad \forall u \in S(\mathbb{R}^2),$$

and, hence operator A extends in continuity to a bounded operator $A: \ H_{s,\alpha}(\mathbb{R}^2) \to H_{s-\alpha}(\mathbb{R}^2)$.

Indeed, putting in lemma 11.2 $\gamma = \alpha$ and noting that $H_{s,0}(\mathbb{R}^2) = H_s(\mathbb{R}^2)$ we obtain the statement required.

Theorem 11.1. *Let operators B_i, $i = 1,2$, have symbols which are homogeneous of order β_i, and*

$$|B_i(\xi_1,\xi_2)| \le c_i|\xi_i|^{\beta_i-\beta_{i1}}\left(|\xi_i|+|\xi_j|\right)^{\beta_{i1}}, \quad \beta_i > \beta_{i1}, \quad i,j=1,2, \quad i \neq j. \tag{11.13}$$

If $s-\alpha+\beta_i+1/2>0$, $i=1,2$, then operators B_i boundedly act from $H_{s-\alpha+\beta_i+1/2}(\mathbb{R}_+)$ into space $H_{s-\alpha,æ-\alpha}(\mathbb{R}^2)$.

Proof. For definiteness let's consider case $i = 1$, $j = 2$. We have $(v_1(x_2) \in C_0^\infty(\mathbb{R}_+))$

$$||B_1(\xi_1,\xi_2)\tilde{v}_1(\xi_2)||^2_{s-\alpha,æ-\alpha} = \int_{\mathbb{R}^2} |B_1(\xi)|^2|\tilde{v}_1(\xi_2)|^2|\xi|^{2(æ-\alpha)}(1+|\xi|)^{2(s-\alpha+æ+\alpha)}d\xi \le$$

(by virtue of (11.13))

$$\le c\int_{\mathbb{R}^2} |\tilde{v}_1(\xi_2)|^2|\xi_1|^{2(\beta_1-\beta_{11})}(|\xi_1|+|\xi_2|)^{2\beta_{11}}|\xi|^{2(æ-\alpha)}(1+|\xi|)^{2(s-æ)}d\xi \le$$

$$\le c\int_{-\infty}^{+\infty}\left(\int_0^{+\infty} |\xi_1|^{2(\beta_1-\beta_{11})}(|\xi_1|+|\xi_2|)^{2\beta_{11}}|\xi|^{2(æ-\alpha)}(1+|\xi|)^{2(s-æ)}d\xi_1\right)|\tilde{v}_1(\xi_2)|^2d\xi_2,$$

and it is left to estimate the inner integral.

Let's represent $(0;+\infty) = (0;1/2]\cup(1/2;+\infty)$.

If $0 < \xi_1 \le 1/2$ then

$$1+|\xi| \sim 1+|\xi_1|+|\xi_2| \sim 1+|\xi_2|,$$

and we have

$$\int_0^{1/2} \xi_1^{2(\beta_1-\beta_{11})}(\xi_1+|\xi_2|)^{2(\beta_{11}+æ-\alpha)}(1+|\xi|)^{2(s-æ)}d\xi_1 \le$$

$$\le (1+|\xi_2|)^{2(s-æ)}\int_0^{1/2} \xi_1^{2(\beta_1-\beta_{11})}(\xi_1+|\xi_2|)^{2(\beta_{11}+æ-\alpha)}d\xi_1 \le$$

(since $\beta_1 > \beta_{11}$)

$$\le c(1+|\xi_2|)^{2(s-æ)}\int_0^{1/2}(\xi_1+|\xi_2|)^{2(\beta_{11}+æ-\alpha)}d\xi_1 \le$$

$$\le c(1+|\xi_2|)^{2(s-\alpha+\beta_1+1/2)}.$$

If $\xi_1 > 1/2$ then

$$|\xi| \sim |\xi_1|+|\xi_2| \sim 1+|\xi|, \quad \text{and}$$

$$\int_{1/2}^{+\infty} \xi_1^{2(\beta_1-\beta_{11})}(|\xi_1|+|\xi_2|)^{2\beta_{11}}|\xi|^{2(æ-\alpha)}(1+|\xi|)^{2(s-æ)}d\xi_1 \leq$$

$$\leq c\int_{1/2}^{+\infty}(1+\xi_1+|\xi_2|)^{2(\beta_1+s-\alpha)}d\xi_1 \leq c(1+|\xi_2|)^{2(s-\alpha+\beta_1+1/2)}$$

if the condition $s-\alpha+\beta_1+1/2<0$ is fulfilled.

So,

$$||B_1(\xi_1,\xi_2)\tilde{v}_1(\xi_2)||^2_{s-\alpha,æ-\alpha} \leq c\int_{-\infty}^{+\infty}|\tilde{v}_1(\xi_2)|^2(1+|\xi_2|)^{2(s-\alpha+\beta_1+1/2)}d\xi_2 \equiv$$

$$\equiv c\,[v_1]^2_{s-\alpha+\beta_1+1/2}. \qquad \blacksquare$$

In order that we can use our previous results everywhere below we will assume that one has perturbation $A^{(\varepsilon)}$ for operator A such that $A^{(\varepsilon)}(\xi)\in C_\alpha$ and $A^{(\varepsilon)}(\xi)$ admits wave factorization with respect to C^a_+, moreover in the limiting case for $\varepsilon\to 0$, $a^2\xi_2^2-\xi_1^2\neq 0$, both symbol $A^{(\varepsilon)}(\xi)$ and elements of wave factorization give the symbol $A(\xi)$ and elements of homogeneous wave factorization. Reasoning as before with perturbed symbol $A^{(\varepsilon)}(\xi)$ and passing to the limit we obtain the system (11.9) - (11.12) again.

Let us denote $\ell_i = s-\alpha+\beta_1+1/2$, $\gamma=\max\{-\beta_1-1/2,-\beta_2-1/2\}$. The following result holds.

Theorem 11.2. *Let $\ell_i<0$, $i=1,2$, A be a pseudodifferential operator with symbol $A(\xi)\in\tilde{C}_\alpha$, æ be the index of homogeneous wave factorization of $A(\xi)$ with respect to C^a_+, symbols $B_i(\xi)$ of pseudodifferential operators B_i be homogeneous of order β_i, $i=1,2$. Let the conditions (11.7) and (11.13) be fulfilled, and*

$$\inf|\det K(\lambda)|\neq 0, \quad \mathrm{Re}\,\lambda=1/2. \tag{11.14}$$

Then there exists unique solution (u,v_1,v_2) of equation (11.4) in space $H_{s,æ}(C^a_+)\times H_{\ell_1}(\mathbb{R}_+)\times H_{\ell_2}(\mathbb{R}_+)$, $æ-s=-1+\delta$, $|\delta|<1/2$, under arbitrary right-hand side $f\in H'_{s-\alpha,æ-\alpha}(C^a_+)$.

A priori estimate holds

$$||u||_{s,æ}+[v_1]_{\ell_1}+[v_2]_{\ell_2}\leq ||f||^+_{\gamma,æ-\alpha}.$$

Proof. Taking into account existence of integrals needed and assumptions above it is left to prove a priori estimate only.

Let us denote the vector with components $\hat{v}_{10}(\lambda)$, $\hat{v}_{11}(\lambda)$, $\hat{v}_{20}(\lambda)$, $\hat{v}_{21}(\lambda)$ by $\hat{V}(\lambda)$, the vector with components $\hat{g}'_{20}(\lambda)$, $\hat{g}'_{21}(\lambda)$, $\hat{g}'_{10}(\lambda)$, $\hat{g}'_{11}(\lambda)$ by $\hat{G}(\lambda)$. System (11.9) - (11.12) in vector notation takes the form

$$\hat{K}(\lambda)\hat{V}(\lambda)=\hat{G}(\lambda).$$

By virtue of (11.14) there exists an inverse matrix $\hat{K}^{-1}(\lambda)$, from which

$$\hat{V}(\lambda) = \hat{K}^{-1}(\lambda)\hat{G}(\lambda).$$

Since the components of matrix $\hat{K}^{-1}(\lambda)$ are bounded (by assumption above) we have

$$||\hat{V}(\lambda)||_0 \le c||\hat{G}(\lambda)||_0,$$

where

$$||\hat{V}(\lambda)||_0^2 = \frac{1}{2\pi i}\int\limits_{1/2-i\infty}^{1/2+i\infty} |\hat{V}(\lambda)|^2 d\lambda,$$

and by virtue of Parceval equality

$$\int\limits_0^{+\infty} |V(x)|^2 dx = \frac{1}{2\pi i}\int\limits_{1/2-i\infty}^{1/2+i\infty} |\hat{V}(\lambda)|^2 d\lambda,$$

we obtain

$$\int\limits_0^{+\infty} |V(x)|^2 dx \le c\int\limits_0^{+\infty} |G(x)|^2 dx. \tag{11.15}$$

Returning to unknowns v_1, v_2 of system (11.8) we have from (11.15)

$$[v_i]_0 \le \left(\left[b_1^{-1}g_1\right]_0 + \left[b_2^{-1}g_2\right]_0\right), \quad i = 1, 2.$$

According to $\ell_i < 0$ we have

$$[v_i]_{\ell_i} \le [v_i]_0,$$

and it is left to estimate $\left[b_i^{-1}g_i\right]_0$, $i = 1, 2$.

It was shown (see proof of Lemma 11.1) ,

$$\operatorname{ord} b_1(\xi_2) = \beta_1 + æ - \alpha + 1.$$

By assumption of Theorem 11.2 we have $s = æ + 1 - \delta$, and since $\ell_i = s - \alpha + \beta_i + 1/2 < 0$, then we have $æ + \beta_1 + 1 - \alpha < \delta - 1/2 < 0$.

Then

$$\left[b_1^{-1}g_1\right]_0^2 \le c\,[g_1]^2_{\alpha-æ-\beta_1-1} \le$$
(by Proposition 3.1.2)

$$\le c||A_{=}^{-1}\ell\tilde{f}||_{\alpha-æ-\beta_1-1/2} \le$$
(by Corollary 11.1)

$$\le c||\ell\tilde{f}||_{-\beta_1-1/2,æ-\alpha} \le c||f||^{+}_{-\beta_1-1/2,æ-\alpha} \le c||f||^{+}_{\gamma,æ-\alpha},$$

because $H_{s_1} \subset H_{s_2}$ if $s_1 < s_2$.

The norm $\left[b_2^{-1}g_2\right]_0$ can be estimated analogously, and we have

$$[v_1]_{\ell_1} + [v_2]_{\ell_2} \le ||f||^+_{\gamma,æ-\alpha}. \tag{11.16}$$

Let us estimate $||u||_{s,æ}$. According to Theorem 8.1.3 we have

$$\tilde{U}(x) = a_{\neq}^{-1}(x)(x_1 - x_2 - i)^{-1}G_2'(x_1 - x_2 - i)a_{=}^{-1}(x)g(x),$$

where

$$\tilde{U}(x) = \tilde{u}\left(\frac{x_2 + x_1}{2}, \frac{x_2 - x_1}{2a}\right),$$

$$a_{\neq}(x) = A_{\neq}\left(\frac{x_2 + x_1}{2}, \frac{x_2 - x_1}{2a}\right), \quad a_{=}(x) = A_{=}\left(\frac{x_2 + x_1}{2}, \frac{x_2 - x_1}{2a}\right),$$

and we must take

$$g(x) = \left(\ell f\right)(x) - B_1(x)v_1(x_2) - B_2(x)v_2(x_1). \tag{11.17}$$

We have

$||u||_{s,æ} = ||U||_{s,æ} = ||a_{\neq}^{-1}(x)(x_1 - x_2 - i)^{-1}G_2'(x_1 - x_2 - i)a_{=}^{-1}(x)g(x)||_{s,æ} \le$
(by Lemma 11.2; ord $a_{\neq}(x) = æ$)

$\le c||(x_1 - x_2 - i)^{-1}G_2'(x_1 - x_2 - i)a_{=}^{-1}(x)g(x)||_{s-æ,0} \le$
(on proposition 4.1.1)

$$\le c||G_2'(x_1 - x_2 - i)a_{=}^{-1}(x)g(x)||_{s-æ-1}.$$

Note that according to our assumptions $æ - s = -1 + \delta$, and hence $s - æ - 1 = -\delta$, $|\delta| < 1/2$. It is well-known (Proposition 4.2.3) that in this case the operator $G_2' : \tilde{H}_{-\delta}(\mathbb{R}^2) \to \tilde{H}_{-\delta}(\mathbb{R}^2)$ is bounded.

Then

$$||u||_{s,æ} \le c||(x_1 - x_2 - i)a_{=}^{-1}(x)g(x)||_{s-æ-1} \le$$

(we use sequentially again Proposition 4.1.1 and Lemma 11.2; ord $a_{=}(x) = \alpha - æ$)

$$\le c||a_{=}^{-1}(x)g(x)||_{s-æ} \le c||g||_{s-\alpha,æ-\alpha} \le$$

$$\le c\left(||\ell f||_{s-\alpha,æ-\alpha} + ||B_1v_1||_{s-\alpha,æ-\alpha} + ||B_2v_2||_{s-\alpha,æ-\alpha}\right)$$

taking into account the formula (11.17).

With help of Theorem 11.1

$$||B_iv_i||_{s-\alpha,æ-\alpha} \le c\,[v_i]_{\ell_i}$$

from where by virtue of estimate (11.16) we obtain finally

$$||u||_{s,æ} \le c||f||^+_{\gamma,æ-\alpha},$$

because $s - \alpha < \gamma$. ∎

Appendix 1: The multidimensional Riemann problem

One can consider this section as a continuation of n° 4.2. Here we use its notations.

By boundary Riemann problem for a half-plane with a parameter we mean (cf. [105]) the problem of finding two functions $\Phi_\pm(x)$ from $L_2(\mathbb{R}^m)$ which admit analytical continuation into half-planes $\mathbb{C}_\pm$ respectively under fixed x', and which are related by a linear equation on $\mathbb{R}^m$

$$\Phi_+ = H\Phi_- + h, \tag{A1.1}$$

where $H(x)$, $h(x)$ are given functions defined on $\mathbb{R}^m$ almost everywhere, moreover $H(x'; -\infty) = H(x'; +\infty)$.

The function $H(x)$ is called the coefficient, and $h(x)$ is called the free term. If $h(x) \equiv 0$ then the problem (A1.1) is called homogeneous Riemann problem.

By index of function $H(x)$ on variable x_m under fixed x' we mean the variation [105] of argument of function $H(x)$ when x_m varies from $-\infty$ to $+\infty$.

We consider the vanishing index case because in multidimensional variant one meets this situation only.

By factorization of function $H(x)$ on variable x_m is meant its representation in form

$$H = H_+ \cdot H_-,$$

where $H_\pm$ admits an analytic continuation into $\mathbb{C}_\pm$ respectively under fixed x'.

If $\operatorname{ind} H = 0$, and H is infinitely differentiable on $\dot{\mathbb{R}}^m$ (it is a one-point compactification of $\mathbb{R}^m$) then this factorization exists and is unique. Besides it is constructed in explicit form with the help of a Cauchy type integral by formulas

$$H_\pm(x', x_m) = \exp\left(\pm \frac{i}{2\pi} \lim_{\tau \to +0} \int_{-\infty}^{+\infty} \frac{\ln H(x', y_m) dy_m}{y_m - x_m - i\tau} \right).$$

In order to state the Riemann problem correctly it is necessary to choose functional spaces in which one seeks the solution. Such spaces we take as $L_2(\mathbb{R}^m) \ni \Phi_\pm$, and we suggest that H is infinitely differentiable on $\dot{\mathbb{R}}^m$, $h(x) \in L_2(\mathbb{R}^m)$.

Proposition A1.1. Under the assumption above the problem (A1.1) has a unique solution which is given by formulas

$$\begin{cases} \Phi_+ = \frac{1}{2} H_+ (I + \Pi^\circ) H_+ h \\ \Phi_- = -\frac{1}{2} H_-^{-1} (I - \Pi^\circ) H_+ h \end{cases} \tag{A1.2}$$

The boundary Riemann problem can be reduced in an equivalent way to a certain so-called characteristic singular integral equation of type (see [105])

$$a(x)\varphi(x) - \frac{b(x)i}{\pi} \text{ v.p.} \int_{-\infty}^{+\infty} \frac{\varphi(x', y_m) dy_m}{y_m - x_m}, \tag{A1.3}$$

where $a(x) = \frac{1}{2}(1 - H(x))$, $b(x) = \frac{1}{2}(1 + H(x))$.

Our goal is to obtain a multidimensional analogue for the boundary Riemann problem and corresponding operators for Π^+, Π^-.

So, we have demonstrated the role in the classic Riemann problem in half-space (or with parameter x') which is played by the Hilbert transform (or operator), i.e., singular integral operator on $\mathbb{R}$:

$$(Su)(x) = \frac{1}{\pi i}\text{v.p.}\int\limits_{-\infty}^{+\infty} \frac{u(t)dt}{t-x} \equiv$$

$$\equiv \frac{1}{\pi i}\lim_{\varepsilon\to 0}\left(\int\limits_{-\infty}^{x-\varepsilon} + \int\limits_{x+\varepsilon}^{+\infty}\right)\frac{u(t)dt}{t-x}, \quad x \in (-\infty; +\infty).$$

Here we have changed the constant so that the operators Π° and S differ with some constant.

The symbol of operator S (sign – function) by virtue of homogeneity of order 0 one can consider defined on the 0 - dimensional unit sphere which is a two-point set (the points ± 1 on real axis), and spectra of Hilbert operator (i.e., values of symbol) consists of two points ± 1 also. We give the following generalization of this spectral property. On the unit sphere S^{m-1} of m - dimensional space $\mathbb{R}^m$, let the sign - function be given, and its values ± 1 taken on two connected components $M_\pm (M_+ \cup M_- = S^{m-1})$. This function will be symbol of operator which is a multidimensional generalization of the Hilbert operator.

For convex cone C^m in m - dimensional space and corresponding radial tube domain $T(C^m)$ (see Chapter 2) we define a Bochner kernel (the difference from [285] is not significant)

$$B_m(z) = \int\limits_{C^m} e^{iz\cdot\xi}d\xi,$$

$z = (z_1, \ldots, z_m)$, $z_j = x_j + iy_j$, $j = 1, 2, \ldots, m$.

Let us give some examples of Bochner kernels [285].

1) $C^m = \{x \in \mathbb{R}^m : \ x_j > 0, \ j = 1, 2, \ldots, m\}$, $\overset{*}{C}{}^m = C^m$,

$$B_m(z) = \frac{i^m}{z_1 z_2 \ldots z_m};$$

2) $C^m = \{x \in \mathbb{R}^m : \ ax_m > |x'|, a > 0\}$,

$$B_m(z) = \frac{a \cdot 2^{m-1} \cdot \pi^{\frac{m-1}{2}} \cdot \Gamma\left(\frac{m}{2}\right)}{\left(z_1^2 + z_2^2 + \cdots + z_{m-1}^2 - a^2 z_m^2\right)^{m/2}}.$$

3) This example has not been included in standard textbooks.

$$C^m = \{x \in \mathbb{R}^m : \ x_m > \sum_{j=1}^{m-1} a_j|x_j|, \ a_j > 0, \ j = 1, \ldots, m-1\}.$$

Calculations give the following:

$$\int\limits_{C^m} e^{i(x_1\xi_1+\cdots+x_{m-1}\xi_{m-1})+iz_m\xi_m}d\xi =$$

$$(z_m = x_m + iy_m,\ y_m > 0)$$

$$= \int\limits_{\mathbb{R}^{m-1}} \left(\int\limits_{\sum_{j=1}^{m-1} a_j|\xi_j|}^{+\infty} e^{iz_m\xi_m}d\xi_m \right) e^{ix'\cdot\xi'}d\xi' =$$

$$= \frac{1}{iz_m}\prod_{j=1}^{m-1}\int\limits_{-\infty}^{+\infty} e^{ix_j\xi_j}e^{ia_j|\xi_j|z_m}d\xi_j =$$

$$= \frac{i^{m-1}a_1\dots a_{m-1}z_m^{m-2}2^{m-1}}{(x_1^2-a_1^2z_m^2)\dots(x_{m-1}^2-a_{m-1}^2z_m^2)},$$

because

$$\int\limits_{-\infty}^{+\infty} e^{ix_j\xi_j}e^{ia_j|\xi_j|z_m}d\xi_j =$$

$$= \int\limits_{-\infty}^{0} e^{i\xi_j(x_j+a_jz_m)}d\xi_j + \int\limits_{0}^{+\infty} e^{i\xi_j(x_j-a_jz_m)}d\xi_j =$$

$$= \frac{1}{i(x_j+a_jz_m)} - \frac{1}{i(x_j-a_jz_m)} = \frac{2a_jz_mi}{x_j^2-a_j^2z_m^2}.$$

These calculations imply that the Bochner kernel has the form in this case:

$$B_m(z) = (2i)^{m-1}z_m^{m-2}\prod_{j=1}^{m-1}\frac{a_j}{z_j^2-a_j^2z_m^2}.$$

Let $\chi(C^m)$ denote the characteristic function of cone C^m, and we take a function $u(\xi)\in S(\mathbb{R}^m)$. Let us consider

$$\int\limits_{\mathbb{R}^m} e^{ix\cdot\xi}\chi(C^m)e^{-y_m\xi_m}u(\xi)d\xi,\quad y_m > 0.$$

As soon as functions $u(\xi)$ and $\chi(C^m)e^{-y_m\xi_m}$ are absolute integrable, then to its product the convolution theorem is applicable, and with the above gives

$$\int\limits_{\mathbb{R}^m} e^{ix\cdot\xi}\chi(C^m)e^{-y_m\xi_m}u(\xi)d\xi =$$
$$= \int\limits_{\mathbb{R}^m} B(x_1-t_1,\dots,x_{m-1}-t_{m-1},z_m-t_m)\tilde{u}(t)dt\ . \tag{A1.4}$$

The limit $\chi(C^m)e^{-y_m\xi_m}u(\xi)$ exists under $y_m \to +0$ in norm of space $L_2(\mathbb{R}^m)$, and consequently there exists a limit in right-hand side of equality (A1.4). Let us introduce an integral operator G_m on functions $u \in S(\mathbb{R}^m)$ by the following formula

$$(G_m u)(x) = 2 \lim_{\tau\to+0} \int_{\mathbb{R}^m} B_m(x_1 - t_1, \ldots, x_{m-1} - t_{m-1}, z_m - t_m)\times$$

$$\times u(t_1, \ldots, t_m)dt - u(x),$$

where one means the limit in L_2 - convergence sense.

Theorem A1.1. *The operator G_m is bounded in space $L_2(\mathbb{R}^m)$, and it is unitary equivalent to a multiplication operator on a function which is equal to 1 in C^m and -1 in $\mathbb{R}^m \setminus C^m$.*

Essentially proof of this theorem follows from previous constructions.

Let us fix $u \in S(\mathbb{R}^m)$ and introduce

$$\Phi(z) = \int_{\mathbb{R}^m} B_m(z-t)u(t)dt, \quad z \in T(\overset{*}{C}{}^m).$$

It is well-known [285] that $\Phi(z)$ is analytic in $T(\overset{*}{C}{}^m)$ and satisfies the estimate

$$\int_{\mathbb{R}^m} |\Phi(x+iy)|^2 dx \le M, \quad \forall y \in \overset{*}{C}{}^m, \tag{A1.5}$$

where the constant M depends on function Φ only.

Let $A_m(\mathbb{R}^m)$ be a space of functions defined almost ewerywhere on $\mathbb{R}^m$, and which admit an analytic continuation into $T(\overset{*}{C}{}^m)$ satisfying condition (A1.5). It is obvious that $A_m(\mathbb{R}^m)$ is a subspace in $L_2(\mathbb{R}^m)$. Let us introduce subspace

$$B_m(\mathbb{R}^m) \equiv L_2(\mathbb{R}^m) \ominus A_m(\mathbb{R}^m)$$

as the orthogonal complement of subspace $A_m(\mathbb{R}^m)$ in $L_2(\mathbb{R}^m)$.

So,

$$A_m(\mathbb{R}^m) \oplus B_m(\mathbb{R}^m) = L_2(\mathbb{R}^m).$$

Now let us formulate an, interesting for us, multidimensional analogue of the so-called jump problem which is well-known in ordinary complex analysis [105,180]: given function $u \in L_2(\mathbb{R}^m)$ finding functions $u_1 \in A_m(\mathbb{R}^m)$, $u_2 \in B_m(\mathbb{R}^m)$ such that the equality

$$u_1(x) + u_2(x) = u(x) \tag{A1.6}$$

holds almost everywhere on $\mathbb{R}^m$. (In the classical jump problem there is sign "$-$" in equality (A1.6) instead of sign "$+$" but it is not essential).

With operator G_m let us introduce two operators

$$P'_+ = \frac{1}{2}(I + G_m),$$

$$P'_- = \frac{1}{2}(I - G_m).$$

Lemma A1.1. *Operator P'_+ is an orthogonal projector from $L_2(\mathbb{R}^m)$ on $A_m(\mathbb{R}^m)$, and P'_- on $B_m(\mathbb{R}^m)$.*

Proof. Actually by definition of operator G_m, taking into account Theorem A1.1, we conclude that

$$\left(P'_+ u\right)(x) = \lim_{y \to 0,\, y \in C^m} \int\limits^{*}_{\mathbb{R}^m} B_m(z-t)u(t)dt \qquad (A1.7)$$

at least for $u \in S(\mathbb{R}^m)$. Extending (A1.7) on full $L_2(\mathbb{R}^m)$ by virtue of the fact that $S(\mathbb{R}^m)$ is dense in $L_2(\mathbb{R}^m)$ and $L_2(C^m)$ is isomorphic to $A_m(\mathbb{R}^m)$ (Corollary 3.3.1) we obtain the first statement of Lemma A1.1. The second is obvious:

$$P'_- = I - P'_+ = I - \frac{1}{2}(I + G_m) = \frac{1}{2}(I - G_m). \quad \blacksquare$$

Remark A1.1. One can consider the statement of Lemma A1.1 as a multidimensional analogue of the Plemelj-Sokhotsky formulas [105,180].

If $u(x) \in S(\mathbb{R})$ then function

$$\Phi(z) = \frac{1}{2\pi i} \int\limits_{-\infty}^{+\infty} \frac{u(t)dt}{t - z}, \qquad z = x + iy,$$

is analytic in the upper and lower half-planes of complex plane $\mathbb{C}$ and it has corresponding boundary values

$$\Phi_+(x) = \frac{1}{2}u(x) + \frac{1}{2\pi i}\text{v.p.} \int\limits_{-\infty}^{+\infty} \frac{u(t)dt}{t - x},$$

$$\Phi_-(x) = -\frac{1}{2}u(x) + \frac{1}{2\pi i}\text{v.p.} \int\limits_{-\infty}^{+\infty} \frac{u(t)dt}{t - x},$$

from which two Plemelj-Sokhotsky formulas follow

$$\Phi_+(x) - \Phi_-(x) = u(x)$$

$$\Phi_+(x) + \Phi_-(x) = (Su)(x),$$

where $(Su)(x)$ denotes $\frac{1}{\pi i}\text{v.p.} \int\limits_{-\infty}^{+\infty} \frac{u(t)dt}{t-x}$.

Clearly the role of projectors P'_+, P'_- above in the one-dimensional case is taken by operators

$$\frac{1}{2}(I + S) \text{ and } \frac{1}{2}(I - S).$$

One property only tends to "obscure" the analogue above: the operator S is defined by "inner" way (on the real line), but the operator G_m is defined by exit into the complex domain. Although similar definitions for operator S were used in the classics [84].

Lemma A1.2. *The jump problem is uniquely solvable for arbitrary right-hand side $u \in L_2(\mathbb{R}^m)$, and*

$$u_1 = P'_+ u, \quad u_2 = P'_- u.$$

Proof. It is enough to establish uniquiness of representation (A1.6) only. Assuming otherwise, we have a pair of functions $u'_1 \in A_m(\mathbb{R}^m)$, $u'_2 \in B_m(\mathbb{R}^m)$ satisfying equality (A1.6). Then

$$u_1(x) - u'_1(x) = u'_2(x) - u_2(x),$$

and $u_1 - u'_1 \in A_m(\mathbb{R}^m)$, $u'_2 - u_2 \in B_m(\mathbb{R}^m)$. By virtue of the fact that one of these spaces is the orthogonal complement to the other we obtain

$$u_1 - u'_1 \equiv 0, \quad u'_2 - u_2 \equiv 0. \quad \blacksquare$$

We pass to the multidimensional Riemann problem in the general statement.

Setting of problem: finding two functions $\Phi_\pm(x)$ such that $\Phi_+ \in A_m(\mathbb{R}^m)$, $\Phi_- \in B_m(\mathbb{R}^m)$, which satisfy linear relation

$$\Phi_+(x) = W(x)\Phi_-(x) + w(x) \qquad (A1.8)$$

almost everywhere on $\mathbb{R}^m$, $W(x)$, $w(x)$ are given functions.

As in the one-dimensional case in dependence on choice of functional spaces which include $W(x)$, $w(x)$ one has different variants of the multidimensional Riemann problem setting. We will keep our focus on the situation when $W(x)$ is homogeneous of order 0 and an infinitely differentiable function, and $w(x)$ is taken from class $L_2(\mathbb{R}^m)$ assuming forthcoming consideration of singular integral equations.

So, let $W(x)$ be homogeneous of order 0, and $W(x) \in C^\infty(\mathbb{R}^m \setminus \{0\})$ (i.e., $W(x) \in C^\infty(S^{m-1})$, where S^{m-1} is a unit sphere in $\mathbb{R}^m$). Everywhere below C^m is a convex sharp cone in $\mathbb{R}^m$.

Definition A1.1. *A homogeneous 0 - wave factorization of function $W(x)$ with respect to C^m is its representation in form*

$$W = W_{\neq} W_{=},$$

where $W^{\pm 1}_{\neq}$ $(W^{\pm 1}_{=})$ admits bounded analytic continuation into $T(\overset{*}{C}{}^m)$ $(T(-\overset{*}{C}{}^m))$, and they are homogeneous of order 0 on variables $x_1 + iy_1, \ldots, x_m + iy_m$, and $W_{\neq}(W_{=}) \in L_\infty(S^{m-1})$.

(The last condition means that $W_{\neq}$ $(W_{=})$ are bounded on S^{m-1} up to a set of measure 0).

Theorem A1.2. *Let $W(x)$ be homogeneous of order 0, $W(x) \in C^\infty(S^{m-1})$, $W(x) \neq 0, \forall x \in S^{m-1}$. If $W(x)$ admits a homogeneous 0 - wave factorization with* *respect to C^m then problem (A1.8) for arbitrary right-hand side $w \in L_2(\mathbb{R}^m)$ has a unique solution which is given by formulas*

$$\begin{cases} \Phi_+ = \frac{1}{2} W_{\neq}(I + G_m) W_{\neq}^{-1} w \\ \Phi_- = -\frac{1}{2} W_{=}^{-1}(I - G_m) W_{\neq}^{-1} w \, . \end{cases} \tag{A1.9}$$

Proof. Taking into account a 0 - wave factorization, the equality (A1.8) can be rewritten in form

$$\Phi_+(x) W_{\neq}^{-1}(x) = \Phi_-(x) W_=(x) + W_{\neq}^{-1}(x) w(x). \tag{A1.10}$$

By virtue of properties of function $W(x)$ and elements of 0 – wave factorization we conclude that $W_{\neq}^{-1} \in L_\infty(S^{m-1})$, and consequently $W_{\neq}^{-1} w \in L_2(\mathbb{R}^m)$. Let us consider two left summands in formula (A1.9). The function $\Phi_+(z) W_{\neq}^{-1}(z)$ is analytic in $T(\overset{*}{C}{}^m)$ and

$$\sup \int_{\mathbb{R}^m} \left| \Phi_+(x+iy) W_{\neq}^{-1}(x+iy) \right|^2 dx, \quad y \in \overset{*}{C}{}^m,$$

is finite because $\Phi_+(z)$ satisfies (A1.5), but $|W_{\neq}(z)|$ is bounded from above and from below by constants. Thus $\Phi_+ W_{\neq}^{-1} \in A_m(\mathbb{R}^m)$.

Let us consider $\Phi_- W_=$. Since $\Phi_- \in B_m(\mathbb{R}^m)$ then obviously $\operatorname{supp} F^{-1}(\Phi_-) \subset \mathbb{R}^m \setminus C^m$. Further as $W_=(x)$ is bounded then $F^{-1}(W_=)$ exists in distribution sense, and on Corollary 2.3.2 $\operatorname{supp} F^{-1}(W_=) \subset (-\overline{C}{}^m)$. It is not difficult by methods in [92] to convince ourselves that functions from $C_0^\infty(\mathbb{R}^m \setminus \overline{C}{}^m)$ (infinitely differentiable function for which their support belongs to $\mathbb{R}^m \setminus \overline{C}{}^m$) are dense in $L_2(\mathbb{R}^m \setminus C^m)$. Approximating $F^{-1}(\Phi_-)$ by functions from class $C_0^\infty(\mathbb{R}^m \setminus \overline{C}{}^m)$ we see that $F^{-1}(\Phi_- W_=)$ can be realized as convolution

$$\int_{\mathbb{R}^m} F^{-1}(\Phi_-)(x-y) F^{-1}(W_=)(y) dy =$$

$$= \int_{-C^m} F^{-1}(\Phi_-)(x-y) F^{-1}(W_=)(y) dy.$$

It is left to note that under $x \in C^m$ we have $x - y \in C^m$, and consequently $F^{-1}(\Phi_-)(x-y) = 0$. Hense $\operatorname{supp} F^{-1}(\Phi_- W_=) \subset \mathbb{R}^m \setminus C^m$, and thus $\Phi_- W_= \in B_m(\mathbb{R}^m)$.

The last term in formula (A1.10) belongs to $L_2(\mathbb{R}^m)$ by virtue of boundedness $W_{\neq}^{-1}$.

So, the problem (A1.10) is reduced to the jump problem (A1.7) with $u_1 = \Phi_+ W_{\neq}^{-1}$, $u_2 = -\Phi_- W_=$, $u = W_{\neq}^{-1} w$.

By Lemma A1.1 this problem has a unique solution which is given by formulas

$$\begin{cases} \Phi_+ W_{\neq}^{-1} = \frac{1}{2}(I + G_m) W_{\neq}^{-1} w \\ \Phi_- W_= = -\frac{1}{2}(I - G_m) W_{\neq}^{-1} w, \end{cases}$$

from which we have formulas (A1.9). ∎

A natural question appears as to how many of homogeneous of order 0 functions admit 0 - wave factorization. Sufficient conditions belong to the following result (see also [283,284]).

Theorem A1.3. *Let $W(x)$ be homogeneous of order 0, $W \in C^{\infty}(S^{m-1})$, $W(x) \neq 0$, $\forall x \in S^{m-1}$, $m \geq 3$. If $\operatorname{supp} F^{-1}(\ln W) \subset [C^m \bigcup(-C^m)]$ then W admits homogeneous 0 - wave factorization with respect to cone C^m.*

Remark A1.2. Existence of homogeneous 0 - wave factorization means existence of a solution for the special multidimensional homogeneous Riemann problem in classes of homogeneous of order 0 functions satisfying definition A1.1.

This type of problem was considered by V.S.Vladimirov [283,284]. Here we briefly recall its simplified formulation because in [283,284] there were classes of distributions: finding pairs of functions $\Phi_{\pm}$ defined almost everywhere on $\mathbb{R}^m$ and which admit analytic continuations into $T(\pm \overset{*}{C}{}^m)$ satisfying estimates

$$\int\limits_{\mathbb{R}^m} |\Phi_{\pm}(x+iy)|^2\, dx \leq M, \quad \forall y \in \overset{*}{C}{}^m,$$

with constant M depending on $\Phi_{\pm}$ only, and such that almost everywhere on $\mathbb{R}^m$ limiting relation holds

$$\Phi_+(x) = W(x)\Phi_-(x) + w(x), \tag{A1.11}$$

where $W(x)$, $w(x)$ are given functions. (If $w(x) \equiv 0$ then the problem is called homogeneous).

The problem (A1.11) was solved by a well-known in one-dimensional case factorization method [105,180,185] of function $W(x)$ under natural condition $W(x) \neq 0$. Thus, to problem (A1.8) factorization method is applied too, but it is easy to see the difference between (A1.11) and (A1.8).

Proof of Theorem A1.3. Let us write

$$W = W_{\neq} W_{=}. \tag{A1.12}$$

We solve an auxiliary problem

$$W^{(1)} = W^{(1)}_{\neq} + W^{(1)}_{=}, \tag{A1.13}$$

where $W^{(1)}_{\neq}$, $W^{(1)}_{=}$ are functions satisfying definition A1.1, $W^{(1)}$ is given function from class $C^{\infty}(S^{m-1})$, and it is homogeneous of order 0.
(The relation (A1.13) is "logarithmation result" of (A1.12)

$$\ln W = \ln W_{\neq} + \ln W_{=}, \tag{A1.14}$$

but passing from (A1.12) to (A1.14) generally speaking is not equivalent.)

The problem (A1.13) is also a special case of the jump problem.

The functions $W^{(1)}_{\neq}$, $W^{(1)}_{=}$ can be considered as distributions which satisfy conditions of Corollary 2.3.2, therefore supports of inverse Fourier transforms of these distributions belong to $\overset{*}{C}{}^m$ and $-\overset{*}{C}{}^m$ respectively. Then the equality in distribution sense

$$F^{-1}\left(W^{(1)}\right) = F^{-1}\left(W_{\neq}^{(1)}\right) + F^{-1}\left(W_{=}^{(1)}\right)$$

is valid.

By virtue of infinite differentiability of $W^{(1)}$ on sphere S^{m-1} the function $F^{-1}\left(W^{(1)}\right)$ is a singular kernel of Calderon-Zygmund [47,48], and its resriction on S^{m-1} is infinite differentiable too. Let us denote this kernel f, and consider two other kernels,

$$f_+ = \chi_+ f, \quad f_- = \chi_- f,$$

where $\chi_\pm(x)$ is a characteristic function of $\pm C^m$. It is evident that f can be written in form

$$f = f_+ + f_-, \tag{A1.15}$$

because $\operatorname{supp} F^{-1}\left(W^{(1)}\right) \subset C^m \bigcup (-C^m)$.

Transfer in (A1.15) to Fourier images in the principal value sense [171,172], i.e., from kernels to symbols, gives a solution of problem (A1.13) because a symbol of the kernel for which its restriction on S^{m-1} is bounded, is also bounded on S^{m-1}.

So, denoting $W_{\neq}^{(1)} = F(f_+)$, $W_{=}^{(1)} = F(f_-)$ we have $W_{\neq}^{(1)}, W_{=}^{(1)} \in L_\infty(S^{m-1})$. Besides the functions $W_{\neq}^{(1)}$, $W_{=}^{(1)}$ admit an analytic continuation into $T(\overset{*}{C}{}^m)$, $T(-\overset{*}{C}{}^m)$ respectively (by the same Corollary 2.3.2).

Thus, the functions $W_{\neq}^{(1)}(z)$, $W_{=}^{(1)}(z)$ belong to classes $H_0(\overset{*}{C}{}^m)$, $H_0(-\overset{*}{C}{}^m)$ respectively and have bounded boundary values. Then such functions $W_{\neq}^{(1)}(z)$, $W_{=}^{(1)}(z)$ are bounded under $z \in T(\pm\overset{*}{C}{}^m)$ (see [288]).

Now exponentiating the equality (A1.13) we obtain the required factorization for function $W(x)$. ■

It is clear by virtue of non-vanishing $W(x)$ on S^{m-1} that all our constructions are valid for $W^{-1}(x)$ too.

Remark A1.3. If $m \geq 3$ then sphere S^{m-1} is simply connected, and any branch of $\ln W$ is defined uniquely.

If $m = 2$ then in addition to assumptions of Theorem A1.2 one needs one additional assumption

$$\int_{S^1} d \arg W = 0,$$

because circumference S^1 is not simply connected.

Now let us try to find an analogue between the multidimensional Riemann problem and corresponding singular integral equations.

If one considers the so-called characteristic singular integral equation [105,180] in space $L_2(\mathbb{R}^m)$ with operator G_m instead of Hilbert transform (singular Cauchy integral)

$$a(x)u(x) + b(x)(G_m u)(x) = f(x), \tag{A1.16}$$

then repeating "classical arguments" it is not difficult to verify that the equation (A1.16) is equivalent to the multidimensional Riemann problem (A1.8), where

$$W(x) = -\frac{a(x) - b(x)}{a(x) + b(x)}, \quad w(x) = \frac{f(x)}{a(x) + b(x)}.$$

Indeed by "Plemelj-Sokhotsky formulas" we have:

$$u(x) = u_+(x) - u_-(x),$$

where

$$\begin{cases} u_+(x) = \frac{1}{2}(I + G_m)u(x), \\ u_-(x) = -\frac{1}{2}(I - G_m)u(x). \end{cases} \tag{A1.17}$$

Then $(G_m)(x) = u_+(x) + u_-(x)$.
Substituting these expressions for $u(x)$ and $(G_m u)(x)$ in (A1.16) we obtain

$$a(x)\,(u_+(x) - u_-(x)) + b(x)\,(u_+(x) + u_-(x)) = f(x),$$

or

$$(a(x) + b(x))\,u_+(x) - (a(x) - b(x))\,u_-(x) = f(x)$$

from which we have finally

$$u_+(x) - \frac{a(x) - b(x)}{a(x) + b(x)} u_-(x) = \frac{f(x)}{a(x) + b(x)}.$$

The phrase "with additional condition $u_\pm(\infty) = 0$" appearing in [105] can be omitted, because in the $L_2(\mathbb{R}^m)$ - case the solution of problem (A1.8) is unique and is given by formulas (A1.17).

Thus, the theory of characteristic equation (A1.16) will be fully analogous to the corresponding one-dimensional case with all following consequences.

The last part of Appendix 1 is needed for reading of Appendix 2 only.

Let us pass to consideration of a multidimensional singular integral equation in space $L_2(C^m)$ with "constant coefficients", i.e., an equation of type

$$au(x) + \int_{C^m} M(x - y)u(y)dy = v(x), \quad x \in C^m, \tag{A1.18}$$

where $M(x)$ is a singular Calderon-Zygmund kernel from class $C^\infty(\mathbb{R}^m \setminus \{0\})$, $a \in \mathbb{C}$. If one introduces operators

$$(A_i u)(x) = a_i u(x) + \int_{\mathbb{R}^m} M_i(x - y)u(y)dy, \quad x \in \mathbb{R}^m, \quad i = 1, 2,$$

where $a_i \in \mathbb{C}$, M_i is a singular Calderon-Zygmund kernel from class $C^\infty(\mathbb{R}^m \setminus \{0\})$ then the equation (A1.18) in the cone is special case of "paired" equation in space $L_2(\mathbb{R}^m)$

$$(A_1 P_+ + A_2 P_-) = f, \tag{A1.19}$$

where P_+ is restriction operator on C^m, i.e.,

$$(P_+)(x) = \begin{cases} u(x), & x \in C^m \\ 0, & x \notin \overline{C^m}, \end{cases}$$

P_- is restriction operator on $\mathbb{R}^m \setminus \overline{C^m}$.

It is not difficult to verify that (A1.18) is equivalent to (A1.19) with $A_2 \equiv I$ (identity operator), $f(x) = v(x)$ on C^m, and $f(x) = 0$ on $\mathbb{R}^m \setminus C^m$ (meaning that unique solvability of (A1.18) implies unique solvability of (A1.19) and vice versa).

Let $\sigma_i(\xi)$ be symbol of operator A_i, $i = 1, 2$. By transfer in (A1.19) to Fourier images according to above we obtain the singular integral equation

$$\frac{\sigma_1(\xi) + \sigma_2(\xi)}{2}\tilde{u}(\xi) + \frac{\sigma_1(\xi) - \sigma_2(\xi)}{2}(G_m\tilde{u})(\xi) = \tilde{f}(\xi). \tag{A1.20}$$

The multidimensional Riemann problem for characteristic equation (A1.20) has form

$$\Phi_+(\xi) = \sigma_2(\xi)\sigma_1^{-1}(\xi)\Phi_-(\xi) + \sigma_1^{-1}(\xi)\tilde{f}(\xi). \tag{A1.21}$$

Solution of equation (A1.20) (or the same of the problem (A1.21)) can be written according to formula (A1.9), where $W_{\neq}(\xi)$, $W_{=}(\xi)$ are factorization elements of function $\sigma_1^{-1}(\xi)\sigma_2(\xi)$, and $\sigma_1^{-1}\tilde{f}$ must be taken as w.

We will formulate the theorem and write out the formula for solution of equation (A1.20) although it is an explicit copy of corresponding one-dimensional formula [105] with corrections on coefficients and operators.

Theorem A1.4. *Let $\sigma_1^{-1}(\xi)\sigma_2(\xi) \in C^\infty(S^{m-1})$, $\sigma_1^{-1}(\xi)\sigma_2(\xi) \neq 0$, $\forall \xi \in S^{m-1}$. If $\sigma_1^{-1}\sigma_2$ admits 0 - wave factorization with respect to C^m then the equation (A1.19) for arbitrary right-hand side $f \in L_2(\mathbb{R}^n)$ has a unique solution $u \in L_2(\mathbb{R}^m)$ for which its Fourier transform is given by formula*

$$\tilde{u}(\xi) = \frac{\sigma_2(\xi) + \sigma_1(\xi)}{2\sigma_1(\xi)\sigma_2(\xi)}\tilde{f}(\xi) - \frac{\sigma_2(\xi) - \sigma_1(\xi)}{2\sigma_1(\xi)\sigma_2(\xi)}W_{\neq}(\xi)\sigma_1(\xi)\times$$

$$\times\left(G_m\left(W_{\neq}^{-1}\sigma_1^{-1}\tilde{f}\right)\right)(\xi),$$

where $W_{\neq}(\xi)$ is factorization element of function $\sigma_1^{-1}(\xi)\sigma_2(\xi)$.

Further we will briefly describe how one can extend the results obtained.

Definition A1.2. *By a wedge of dimension n in m - dimensional space $\mathbb{R}^m$ we'll mean set of type $\mathbb{R}^n \times C^{m-n}$. By definition we put*

$$\mathbb{R}^0 \times C^m \equiv C^m, \quad \mathbb{R}^m \times C^0 \equiv \mathbb{R}^m.$$

Equations with constant coefficients of (A1.18) type in cone C^m ($n = 0$) were considered above, and in space $\mathbb{R}^m$ ($n = m$) such equations have been investigated for a long time [171]. Here we are interested in the intermediate case $0 < n \leq m-1$.

Let us briefly focus on case $n = m - 1$. Then C^1 is half-axis, and the wedge $\mathbb{R}^n \times C^{m-n}$ transforms into half-space $\mathbb{R}^{m-1} \times (0; +\infty) \equiv \mathbb{R}^m_+$. The equation (A1.18) takes the form respectively

$$au(x) + \int\limits_{\mathbb{R}^m_+} M(x-y)u(y)dy = v(x), \quad x \in \mathbb{R}^m_+. \qquad (A1.22)$$

Invertibility of the left-hand side operator of equation (A1.22) is studied in detail in different spaces by consideration of paired equation of (A1.19) type and following application of Fourier transform [230]. In result it reduces to the classical boundary Riemann problem for upper and lower half-planes on variable ξ_m with $(m-1)$ - dimensional parameter $\xi' = (\xi_1, \ldots, \xi_{m-1})$ (or, one can say, to the corresponding characteristic singular integral equation).

Thus, it is left to consider the case $0 < n \leq m-2$. Certainly, it is clear that the situation here will be analogous to above, and we will encounter a multidimensional Riemann problem with a parameter. So we will define operators G_{m-n} only and formulate the needed result in the following.

We put

$$(G_{m-n}u)(x) = 2 \lim_{\tau \to +0} \int\limits_{\mathbb{R}^{m-n}} B_{m-n}(x_{n+1} - t_{n+1}, \ldots, x_{m-1} - t_{m-1},$$

$$z_m - t_m)u(x_1, \ldots, x_n, t_{n+1}, \ldots, t_m)dt_{n+1} \ldots dt_m - u(x),$$

where $z_m = x_m + i\tau$, $B_{m-n}(z_{n+1}, \ldots, z_m)$ is the Bochner kernel for C^{m-n}.

The equation (A1.18) in this case has the form

$$au(x) + \int\limits_{\mathbb{R}^n \times C^{m-n}} M(x-y)u(y)dy = v(x), \quad x \in \mathbb{R}^n \times C^{m-n}. \qquad (A1.23)$$

Let $W(x)$ be homogeneous of order 0, and $W(x) \in C^\infty(S^{m-1})$.

Definition A1.3. *A homogeneous n - wave factorization of function $W(x)$ with respect to C^{m-n} is its representation in form*

$$W = W_{\neq} W_{=},$$

where $W_{\neq}^{\pm 1}$ $(W_{=}^{\pm 1})$ admits bounded analytic continuation into $T(\overset{*}{C}^{m-n})$ $\left(T(-\overset{*}{C}^{m-n})\right)$ on variables $x_{n+1}, \ldots, x_m$ for almost all $x_1, \ldots, x_n$, it is homogeneous of order 0 on variables $x_1, \ldots, x_n, x_{n+1} + iy_{n+1}, \ldots, x_m + iy_m$, and $W_{\neq}(W_{=}) \in L_\infty(S^{m-1})$.

Theorem A1.5. *Let $\sigma_1^{-1}(\xi)\sigma_2(\xi) \in C^\infty(S^{m-1})$, $\sigma_1^{-1}(\xi)\sigma_2(\xi) \neq 0$, $\forall \xi \in S^{m-1}$. If $\sigma_1^{-1}\sigma_2$ admits n - wave factorization with respect to C^{m-n} then "paired" equation for (A1.23) under an arbitrary right-hand side $v \in L_2(\mathbb{R}^m)$ has unique solution $u \in L_2(\mathbb{R}^m)$ for which its Fourier transform is given by formula*

$$\tilde{u}(\xi) = \frac{\sigma_2(\xi) + \sigma_1(\xi)}{2\sigma_1(\xi)\sigma_2(\xi)} \tilde{v}(\xi) +$$

$$+\frac{\sigma_2(\xi)-\sigma_1(\xi)}{2\sigma_1(\xi)\sigma_2(\xi)}W_{\neq}(\xi)\sigma_1(\xi)\left(G_{m-n}\left(W_{\neq}^{-1}\sigma_1^{-1}\tilde{v}\right)\right)(\xi),$$

where $W_{\neq}(\xi)$ is an element of n - wave factorization for $\sigma_1^{-1}(\xi)\sigma_2(\xi)$.

Appendix 2: Symbolic calculus, Noether property, index, regularization

This appendix is devoted to development of symbolic calculus for multidimensional singular integral operators in non-smooth domains. It does not relate to our basic text, nevertheless we include it in this book because these questions are important.

First we give some information on which we will construct symbolic calculus and index theory. Basic for construction of symbolic calculus will be the local principle which we briefly describe [107,138,114,121,108,139,230].

Let A be a linear bounded operator acting in Banach space B, $A: B \to B$.

We say operator A admits left regularization if there is a linear bounded operator $R_\ell: B \to B$ such that

$$R_\ell A = I + T_1$$

(Everywhere in this section the letter T with indexes or without will denote a compact operator).

Operator R_ℓ is called a left regularizer of operator A.

Analogously, operator A admits right regularization if there is a linear bounded operator $R_r: B \to B$ such that

$$AR_r = I + T_2,$$

and operator R_r in this case is called a right regularizer of operator A.

Proposition A2.1. *If there exist both operator R_ℓ and R_r then their difference is a compact operator.*

Proposition A2.2. *If left regularizator R_ℓ of operator A exists, then operator $R_\ell + T$ is also a left regularizer of operator A.*

We say operator A admits two-side regularization if it admits right and left regularization at the same time. From Propositions A2.1, A2.2 one concludes that if operator A admits two-side regularization then $R_\ell = R_r$.

Solutions of equation $Au = 0$ are called nulls of operator A. A set of these nulls generates a subspace of space B which is called the kernel of operator A, and it is denoted by $\ker A$.

As usually, A^* denotes an operator adjoint to A. The difference

$$\dim\ker A - \dim\ker A^*$$

is called analytical index of operator A, and it is denoted by $\operatorname{Ind} A$. It is obvious $\operatorname{Ind} A$ is finite if and only if both dimensions $\dim\ker A$ and $\dim\ker A^*$ are finite.

If Ind A is finite then operator A is called a Noether operator (or we say, A has Noether property).

Proposition A2.3. *Operator A has Noether property if and only if it admits two-side regularization.*

Let's note the following important properties of Noether operators.

Proposition A2.4. *If $A_i: B \to B$, $i = 1, 2$ are Noether operators, then both operators $A_1 A_2$ and $A_2 A_1$ have Noether property, and*

$$\operatorname{Ind}(A_1 A_2) = \operatorname{Ind}(A_2 A_1) = \operatorname{Ind} A_1 + \operatorname{Ind} A_2 .$$

The conclusion of Proposition A2.4 is called sometimes "logarithmic property of index".

Proposition A2.5. *If operator A has Noether property, then operator $A + T$ has Noether property also, and*

$$\operatorname{Ind}(A + T) = \operatorname{Ind} A.$$

Proposition A2.6. *If operator A has Noether property, then there is $\varepsilon > 0$ such that operator $A + A_0$ has Noether property for all linear bounded operators $A_0: B \to B$, $||A_0||_{B \to B} < \varepsilon$.*

Conclusions of Propositions A2.5, A2.6 are called briefly "stability property of index with respect to compact and small perturbations".

Proposition A2.7. *Index* Ind A *of Noether operator A is homotopic invariant, namely if A_t continuously depends on $t \in [0; 1]$, and A_t is Noether operator for all t, then* Ind A_t *does not depend on t.*

(One means continuity in operator topology).

Let us now introduce some concepts related to local principle in form which is needed for the following application.

By essential norm of linear bounded operator $A: L_2(D) \to L_2(\mathcal{D})$ we mean the quantity

$$|||A||| = \inf ||A - T||,$$

where T runs on the set of all compact operators.

If M is a measurable set in D, then P_M is a projector on set M, i.e.,

$$(P_M f)(x) = \begin{cases} f(x), & x \in M \\ 0, & x \notin M . \end{cases}$$

Operator A is called an operator of local type if for any two closed non-intersecting sets F_1 and F_2 from $\overline{D}$ the operator $P_{F_1} A P_{F_2}$ is compact.

Let $A_1, A_2: L_2(D) \to L_2(D)$ be two linear bounded operators. We will call them equivalent in point $x_0 \in D$ if for any $\varepsilon > 0$ there is a neighbourhood $u \ni x_0$ such that

$$|||A_1 P_u - A_2 P_u||| < \varepsilon \quad \text{and} \quad |||P_u A_1 - P_u A_2||| < \varepsilon .$$

When A_1, A_2 are operators of local type there is sufficient validity of at least one inequality.

Let $G \subset \mathbb{R}^m$ be another domain, φ be a homeomorphic measurable mapping of some neighbourhood $u \subset D$ on $v \subset G$.

By definition φ does not distort the measure if for an arbitrary measurable set $E \subset u$ the inequality holds:

$$c_1 \operatorname{mes} \varphi(E) \leq \operatorname{mes} E \leq c_2 \operatorname{mes} \varphi(E),$$

where c_1, c_2 are positive constants not depending on E.

Let us denote by T_φ the operator which for every function $f \in L_2(G)$ vanishing out v, is associated the function T_φ by rule

$$(T_\varphi f)(x) = \begin{cases} f(\varphi(x)), & x \in u \\ 0, & x \notin u. \end{cases}$$

In case when φ does not distort the measure, T_φ is a linear bounded operator acting from $L_2(v)$ into $L_2(u)$.

Let us now take two linear bounded operators

$$A_1 : L_2(D) \to L_2(D) \text{ and } A_2 : L_2(G) \to L_2(G).$$

We say operator A_1 in point $x_0 \in D$ is quasiequivalent to operator A_2 at point $y_0 \in G$ if there are two neighbourhoods $u \ni x_0$ and $v \ni y_0$ and a homeomorphic non-distorting of measure transformation φ of neighbourhood u on v such that $y_0 = \varphi(x_0)$ and the operators $T_{\varphi^{-1}} P_u A_1 P_u T_\varphi P_v$, $P_v A_2 P_v$ are equivalent at point y_0.

By a local representative in point $x_0 \in D$ of linear bounded operator $A : L_2(D) \to L_2(D)$ of local type we mean an arbitrary linear bounded operator $A' : L_2(G) \to L_2(G)$ which in point $y_0 \in G$ is quasiequivalent to operator A under local homeomorphism φ non-distorting of measure and such that $\varphi(x_0) = y_0$.

The local principle declares an equivalence of Noether property of operator $A : L_2(D) \to L_2(D)$ of local type and Noether property of all its local representatives A_x, $x \in \overline{D}$.

Clearly that number of local representatives may be arbitrary, but at the same time they all either have Noether property or not. In that case when these local representatives are homogeneous operators, then existence of Noether property is equivalent to invertibility. Let us recall what are these operators.

Let $A : L_2(\mathbb{R}^m) \to L_2(\mathbb{R}^m)$ be a linear bounded operator.

Let us consider $(\alpha * f)(x) \equiv f(\alpha x)$, $\alpha \in \mathbb{R}$, $\alpha \neq 0$. Operator A is called homogeneous if for arbitrary $\alpha > 0$

$$A(\alpha * f) = \alpha * Af.$$

Finally, the last item needed in this section, some elementary information from algebraic topology.

Let X, Y be topological spaces, $F : X \to Y$ be an arbitrary continuous mapping.

A family of continuous mappings $h_t : X \to Y, 0 \leq t \leq 1$, for which their indices are real numbers $t \in [0;1]$ is called a homotopy if the mapping $H : X \times [0;1] \to Y$ defined by formula

$$H(x,t) = h_t(x), \quad x \in X, \quad t \in [0;1],$$

is continuous. The mappings h_0 and h_1 are called initial and final mappings of homotopy h_t. If the mapping f is the initial mapping of homotopy h_t, then we say also that homotopy h_t is a homotopy of mapping f.

Homotopy $h_t : A \to Y, 0 \leq t \leq 1$, A is some subspace of space X, is called partial homotopy of mapping f if $f\big|_A = h_0$ ($f\big|_A$ denotes restriction of mapping f on A). Homotopy $g_t : X \to Y$ is called extension of homotopy $h_t : A \to Y$ if $g_t\big|_A = h_t$ for all $t \in [0;1]$.

We say the subspace A satisfies the homotopy extension axiom with respect to space Y if every partial homotopy

$$h_t : A \to Y, \quad 0 \leq t \leq 1,$$

of arbitrary continuous mapping $f : X \to Y$ admits an extension $g_t : X \to Y$, $0 \leq t \leq 1$, for which $g_0 = f$.

Proposition A2.8. *Every closed subspace A of an arbitrary binormal space X satisfies a homotopy extension axiom with respect to every finite-triangulable space Y.*

(We recall here that the space X is called binormal if the space $X \times [0;1]$ is normal).

Now we pass to construction of symbolic calculus for Calderon-Zygmund singular integral operators in non-smooth domain. We will start from a description of domain D.

Let us denote D a bounded domain with piece-wise smooth boundary ∂D for which we will assume the following. Let us choose on ∂D the sets $T_0, T_1, \ldots, T_{m-2}$ of points, lines and surfaces such that every T_n is represenred as $T_n = \bigcup\limits_{k=1}^{k_n} T_{nk}$, and for every point $t_{nk} \in T_{nk}$ there is a neighbourhood U_{nk} in $\mathbb{R}^m$ such that $U_{nk} \bigcap D$ is diffeomorphic to cone $C^m_{t_{nk}}$ (in the following called tangent cone) which is obtained from cone $\mathbb{R}^n \times C^{m-n}$ by rotation and parallel transfer, so the origin transforms into point t_{nk}. Under this diffeomorphism $U_{nk} \bigcap \partial D$ must correspond to $\partial C^m_{t_{nk}}$, the point t_{nk} has its place, and the Jacobian of this diffeomorphism in point t_{nk} is equal to 1; it is assumed that for every T_{nk} the cone C^{m-n} is the same (in the following under $C^m_{t_{nk}}$ we will mean the cone for which its top is point 0), C^{m-n} is a convex connected cone not including the whole straight line.

A Calderon-Zygmund operator in domain D is by definition the operator given by formula

$$(Ku)(x) = a(x)u(x) + \int\limits_D K(x, x-y)u(y)dy, \quad x \in D,$$

where $a(x) \in C^\infty\left(\overline{D}\right)$, $\overline{D}$ is closure of D, $K(x,y) \in C^\infty\left(\overline{D} \times (\mathbb{R}^m \setminus \{0\})\right)$,

$K(x,y)$ is homogeneous of order $-m$ on variable y for all x, and

$$\int\limits_{S^{m-1}} K(x,y)dS_y = 0, \quad \forall x \in \overline{D}.$$

A long series of papers of A.P.Calderon and A.Zygmund [47,48,51] is devoted to the study of conditions for boundedness of operator

$$u(x) \longmapsto \int\limits_{\mathbb{R}^m} K(x, x-y)u(y)dy$$

in $L_p(\mathbb{R}^m)$ - spaces. Simplest variants of the Calderon-Zygmund theorem permits us to declare that under our assumptions operator K is bounded in space $L_2(D)$.

With operator K we associate the following types of operators which depend on parameter x_0 :

$$u(x) \longmapsto a(x_0)u(x) + \int\limits_{\mathbb{R}^m} K(x_0, x-y)u(y)dy, \quad x \in \mathbb{R}^m, \qquad (A2.1)$$

$x_0 \in \overset{\circ}{D}$ (interiority of D);

$$u(x) \longmapsto a(x_0)u(x) + \int\limits_{\mathbb{R}^m_{+,x_0}} K(x_0, x-y)u(y)dy, \quad x \in \mathbb{R}^m_{+,x_0}, \qquad (A2.2)$$

$x_0 \in \partial D \setminus \left(\bigcup\limits_{n=0}^{m-2} T_n\right)$ is a boundary point of smoothness, $\mathbb{R}^m_{+,x_0}$ is the half-space generated by tangent plane to ∂D at point x_0 and inner normal which are parallel transferred into the origin;

$$u(x) \longmapsto a(x_0)u(x) + \int\limits_{\mathbb{R}^n \times C^{m-n}_{x_0}} K(x_0, x-y)u(y)dy, \quad x \in \mathbb{R}^n \times C^{m-n}_{x_0}, \quad (A2.3)$$

$x_0 \in T_n$, $n = 0, 1, \ldots, m-2$; $C^{m-n}_{x_0}$ is tangent cone to ∂D in point x_0 which is parallel transferred into origin also.

Lemma A2.1. *Operators (A2.1), (A2.2), (A2.3) are local representatives of operator K in corresponding points of $\overline{D}$.*

Proof. Quasiequivalence of operator K to operators (A2.1), (A2.2) in corresponding points was proved earlier [230] therefore we'll stay on operators of type (A2.3) only.

Let $x_0 \in T_k$, $k = \min\{n\}$, $0 \le n \le m-2$, $T_n \bigcap D_{x_0} \ne \emptyset$, where D_{x_0} is intersection of the ball with center in x_0 of radius $\varepsilon > 0$ with domain D.

According to our assumptions on structure of sets T_n, $0 \le n \le m-2$, there is a diffeomorphism of neighbourhood D_{x_0} on neighbourhood $\tilde{D}_{x_0} \bigcap (\mathbb{R}^k \times C^{m-k}_{x_0})$ having specific properties (see above). Let us denote this diffeomorphism by φ. Let $x, y \in D_{x_0}$. Denote $\varphi(x) = \tau$, $\varphi(y) = t$, $\tau, t \in \tilde{D}_{x_0} \bigcap \left(\mathbb{R}^k \times C^{m-k}_{x_0}\right)$.

Let us consider

$$\int\limits_{D_{x_0}} K(x, x-y)u(y)dy, \quad x \in D_{x_0}. \tag{A2.4}$$

Change of variables in integral starting consideration from kernel:

$$K(x, x-y) = K(\varphi^{-1}(\tau), \varphi^{-1}(\tau) - \varphi^{-1}(t)) =$$

(apply to mapping φ^{-1} Taylor formula and take remainder in integral form)

$$= K\left(\varphi^{-1}(\tau), \left(\varphi^{-1}\right)'(\tau)(\tau - t) + R_1(\tau, t)\right), \tag{A2.5}$$

where $\left(\varphi^{-1}\right)'(\tau)$ is derivative of mapping φ^{-1} in point τ.

Note that since integral (A2.4) is singular, then formula (A2.5) can be applied outside of ε - neighbourhood of point x, and then one must pass to the limit for $\varepsilon \to 0$.

Further, applying Taylor's formula to kernel $K(x, y)$ (outside of ε - neighbourhood of point x also) and taking the remainder in integral form again we obtain

$$K\left(\varphi^{-1}(\tau), \left(\varphi^{-1}\right)'(\tau)(\tau - t) + R_1(\tau, t)\right) =$$

$$= K\left(\varphi^{-1}(\tau), \left(\varphi^{-1}\right)'(\tau)(\tau - t)\right) + R_K(\tau, t)R_1(\tau, t).$$

It is not diifficult to note that smoothness of function $R_1(\tau, t)$ on variables τ, t is stipulated by smoothness of mapping φ, and smoothness of $R_K(\tau, t)$ is stipulated by smoothness of kernel K. In a neighbourhood of point τ these functions satisfy estimate:

$$R_1(\tau, t) = O\left(|\tau - t|^2\right),$$

$$R_K(\tau, t) = O\left(\left|\left(\varphi^{-1}\right)'(\tau)(\tau - t)\right|^{-m-1}\right).$$

If one takes into account that

$$c_1|\tau - t| \leq |\left(\varphi^{-1}\right)'(\tau)(\tau - t)| \leq c_2|\tau - t|$$

(with positive constants c_1, c_2 not depending on τ, t) in a neighbourhood of point $\tau = t$, then we have

$$|R_K(\tau, t)R_1(\tau, t)| \leq c|\tau - t|^{1-m},$$

in other words the kernel $R_K(\tau, t)R_1(\tau, t)$ generates an integral potential type operator (with weak singularity) which as well-known [171] is compact in the space scale under consideration.

Integral (A2.4) after change of variables takes the form:

$$\int\limits_{D_{x_0}} K(x, x-y)u(y)dy =$$

$$= \int\limits_{\tilde{D}_{x_0}\bigcap(\mathbb{R}^k\times C_{x_0}^{m-k})} \tilde{K}(\tau,\tau-t)J(t)u_1(t)dt + \int\limits_{\tilde{D}_{x_0}\bigcap(\mathbb{R}^k\times C_{x_0}^{m-k})} R(\tau,t)J(t)u_1(t)dt,$$

$$\tau\in\tilde{D}_{x_0}\bigcap\left(\mathbb{R}^k\times C_{x_0}^{m-k}\right),$$

where $\tilde{K}(\tau,\tau-t)=K\left(\varphi^{-1}(\tau)\right),\left(\varphi^{-1}\right)'(\tau)(\tau-t)),\ R(\tau,t)=R_K(\tau,t)R_1(\tau,t)$, $J(t)$ is the Jacobian of mapping φ^{-1} in point t.

It is left to recall only that the point x_0 under mapping φ does not change its place, the Jacobian at this point is equal to 1, and we have the quasiequivalence required. ■

Let us turn to operators (A2.1) – (A2.3) and note that they are all homogeneous (since in definition of homogeneity the operator has been acting in $\mathbb{R}^m$ then for operators (A2.2), (A2.3) one means corresponding "paired" operators which are equivalent to (A2.2), (A2.3) in Noetherian and invertibility sense; see above).

According to what has been said above operators (A2.1) – (A2.3) in Fourier images (up to non-principal variations non-affecting on invertibility) have the following operator form:

$$\tilde{u}(\xi)\longmapsto\sigma(x,\xi)\tilde{u}(\xi),\quad x\in\overset{\circ}{D},\tag{A2.6}$$

where $\sigma(x,\xi)$ is symbol of operator K :

$$\sigma(x,\xi)=\lim_{\varepsilon\to0,N\to\infty}\int\limits_{\varepsilon<|y|<N}K(x,y)e^{-iy\cdot\xi}dy,$$

$$\tilde{u}(\xi)\longmapsto\frac{1+\sigma(x,\Theta_x\xi)}{2}\tilde{u}(\xi)+\frac{1-\sigma(x,\Theta_x\xi)}{2}(H\tilde{u})(\xi),\tag{A2.7}$$

$x\in\partial D\setminus\left(\bigcup\limits_{n=0}^{m-2}T_n\right)$, Θ_x is orthogonal transform of $\mathbb{R}^m$ (rotation) transferring hyperplane $x_m=0$ into tangent plane to ∂D in point x, H is Hilbert transform (singular Cauchy integral on variable ξ_m):

$$(H\tilde{u})(\xi)=\frac{1}{\pi i}\ \text{v.p.}\int\limits_{-\infty}^{+\infty}\frac{\tilde{u}(\xi',\tau)d\tau}{\xi_m-\tau};$$

$$\tilde{u}(\xi)\longmapsto\frac{1+\sigma(x,\varphi_{x,n}\xi)}{2}\tilde{u}(\xi)-\frac{1-\sigma(x,\varphi_{x,n}\xi)}{2}\left(G_{m-n}\tilde{u}\right)(\xi),\tag{A2.8}$$

$x\in T_n$, $\varphi_{x,n}$ is rotation of $\mathbb{R}^m$ transferring C^{m-n} into tangent cone C_x^{m-n}.

The formulas (A2.7), (A2.8) appear as a result of well-known variations of symbol under linear change of variables (see [171,172]).

On operator K we will construct operator-function $K(x)$ by formulas (A2.6) - (A2.8) which is defined in $\overline{D}$ and takes its values in set $\mathcal{L}$ of linear bounded operators acting in space $L_2(\mathbb{R}^m)$. We will say operator-function $K(x)$ is invertible if there is an operator-function $K^{-1}(x):\ \overline{D}\to\mathcal{L}$ such that

$$K(x)K^{-1}(x)=K^{-1}(x)K(x)=I,\quad\forall x\in\overline{D}.$$

Operator-function $K(x)$ constructed by formulas (A2.6) - (A2.8) is called operator symbol of operator K.

Theorem A2.1. *Operator K has Noether property in space $L_2(D)$ if and only if its operator symbol is invertible, and*

$$\inf \|K(x)\|_{\mathcal{L}} > 0, \quad x \in \overline{D}. \tag{A2.9}$$

Proof. Indeed, Noether property of operator K is equivalent to Noether property (and by virtue of homogeneity it is equivalent to invertibility) of its local representatives according to local principle. The condition (A2.9) is a consequence of the fact that separate types of local representatives are defined on non-compact sets although local principle is valid for compacts only.

Let us verify that if $\inf \|K(x)\|_{\mathcal{L}} = 0$ then the operator can not have Noether property. Really, in this case there is a sequence $\{x_n\}_{n=1}^{\infty}$, $x_n \in \overline{D}$ such that $\lim_{n\to\infty} \|K(x_n)\| = 0$. Since the sequence $\{x_n\}_{n=1}^{\infty}$ is bounded then from it one can choose convergent subsequence $\{x_{n_k}\}_{k=1}^{\infty}$ such that $\lim_{k\to\infty} x_{n_k} = x$, $x \in \overline{D}$. Let us choose now a sequence $\{y_k\}_{k=1}^{\infty}$, $y_k \in \overset{\circ}{D}$ such that $\lim_{n\to\infty} (y_{n_k} - x_{n_k}) = 0$. Then evidently

$$\lim_{k\to\infty} K(y_k) = K(x) = 0,$$

besides $K(x)$ has the form (A2.6) by virtue of smoothness property of symbol $\sigma(x,\xi)$. Since (A2.6) is a multiplication operator on $\sigma(x,\xi)$ then from this follows that

$$\sigma(x,\xi) = 0.$$

The last means operator K has no local Noether property [230], and therefore it has no Noether property. ∎

Let us observe some properties of operator-function $K(x)$.

Theorem A2.2. *The operator-function $K(x)$ constructed on operator K is continuous in $\overset{\circ}{D}$, on $\partial D \setminus \left(\bigcup_{n=0}^{m-2} T_n \right)$, on every component T_n, $n = 0, 1, \ldots, m-2$.*

This is consequence of representation (A2.6) - (A2.8) and smoothness properties of symbol $\sigma(x,\xi)$.

Theorem A2.3. Algebra $\mathcal{A}$ generated by operator of K type and factorized on ideal of compact operators is isomorphic to algebra $\mathcal{A}'$ of operator-function of form (A2.6) - (A2.8).

This is a consequence of the local equivalence property: local representatives of product and sum of local type operators are a product and sum of their local representatives.

Theorem A2.4. Operators $A_1, A_2 \in \mathcal{A}$ are homotopic in $\mathcal{A}$ if and only if their operator symbols are homotopic in $\mathcal{A}'$.

We will not dwell on the proof of this sentence: it is absolutely elementary.

The statement of theorem A2.3 is strictly speaking symbolic calculus. It looks in general as though it may be tautological, but at least one can extract from it some result on Noether property of operator K.

Let x be a smoothness point of boundary ∂D, $m \geq 3$. On $\sigma(x,\xi)$ let us define the local topological characterictic d_x as divided on π increment of argument of function $\sigma(x,\xi)$ under variation of ξ from "south pole" $-\overline{n}_x$ to "north pole" $\overline{n}_x$ on sphere S^{m-1} along any arc of large half-circumference, where $\overline{n}_x$ is unit inner normal vector in point x to boundary ∂D. Let us denote

$$T = \left\{ d_x \,:\, x \in \partial D \setminus \left(\bigcup_{n=0}^{m-2} T_n \right) \right\}.$$

If $m = 2$ let us define $d_x^{\pm}$ as divided on π increments of the argument of function $\sigma(x,\xi)$ under variation of ξ from "south" to "north" polė along "left" (minus) or "right" (plus) half-circumferences.

Theorem A2.5. If the following conditions are satisfied:

1) $\sigma(x,\xi) \neq 0, \quad \forall x \in \overline{D}, \quad \xi \in S^{m-1}$;

2) $\overline{T} \subset (-1;1)$;

3) $\sigma(x,\xi)$ admits n - wave factorization for $\forall x \in T_n$, $n = 0,1,\ldots,m-2$, with respect to cone C_x^{m-n}, then operator K has Noether property in space $L_2(D)$.

Remark A2.1. Since set T_n generally speaking is non-closed, then if $x \in \overline{T}_n$ it may be that $x \notin T_n$. In this case one needs existence of n - wave factorization with respect to cone C_x^{m-n} which is obtained as the limiting state of cone C_y^{m-n}, $y \in T_n$, for $y \to x$.

Before proceeding to discussion of the proof for Theorem A2.5, let us give an equivalent formulation of condition 2) of this therem.

Let γ be a continuous closed curve on the complex plane not including the origin. By index of curve γ (notation $\operatorname{ind}\gamma$) we mean number of turns of γ around the origin.

Let again x be a smoothness point of boundary ∂D. In case $m \geq 3$ let us define curve γ_x in the following way. Join the points $\sigma(x,-\overline{n}_x)$ and $\sigma(x,\overline{n}_x)$ on the complex plane by a line segment and add to it the values $\sigma(x,\xi)$ when ξ variates from $-\overline{n}_x$ to $\overline{n}_x$. As a result we will obtain the closed curve γ_x mentioned. In case $m = 2$ we construct two curves $\gamma_x^{\pm}$ depending on whether ξ variates along the left or right half-circumference.

Let us define such curves γ_x for every point of boundary ∂D, for the points $x \in T_n$, $n = 0,1,\ldots,m-2$, meaning under γ_x the "fan" of curves related to all possible limiting values of $\overline{n}_y$, $y \to x$ (y is a smoothness point). Then the condition 2) of Theorem A2.5 can be formulated in the following way:

$$2') \qquad \begin{array}{ll} 0 \notin \gamma_x, \ \operatorname{ind}\gamma_x = 0, & (m \geq 3) \\ 0 \notin \gamma_x^{\pm}, \ \operatorname{ind}\gamma_x^{\pm} = 0, & (m = 2) \end{array}$$

for all $x \in \partial D$.

We give a full formulation also of an equivalent to Theorem A2.5 for which its proof is the proof of Theorem A2.5.

Theorem A2.5′. *If the following conditions satisfied:*

1) $\sigma(x,\xi) \neq 0, \ \forall x \in \overline{D}, \quad \xi \in S^{m-1}$;

2) $0 \notin \gamma_x, \ \operatorname{ind}\gamma_x = 0 \ (m \geq 3); \ \ 0 \notin \gamma_x^{\pm}, \ \operatorname{ind}\gamma_x^{\pm} = 0 \ (m = 2)$ *for all* $x \in \partial D$;

3) $\sigma(x,\xi)$ admits n - wave factorization for $\forall x \in \overline{T}_n$, $n = 0, 1, \ldots, m-2$, with respect to cone C_x^{m-n},
then the operator K has Noether property in space $L_2(D)$.

Proof. It is necessary to observe that the conditions 1) — 3) represent conditions for invertibility of operators (A2.6) — (A2.8) which are roughly speaking local representatives of operator K. ∎

Now we will start to calculate the index of Noether operator K.

As it follows from Propositions A2.3 — A2.5 homotopies, small perturbations and perturbations by compact summand don not change the index of Noether operator. Since the correspondence between operators K and operator-functions $K(x)$ up to compact operators is one-to-one (Theorem A2.3) then we will simplify operator-function $K(x)$ maximally using homotopies (Theorem A2.4) and small perturbations.

Definition A2.1. *Symbol $\sigma(x,\xi)$ is called elliptic if $\sigma(x,\xi) \neq 0$, $\forall x \in \overline{D}$, $\xi \in S^{m-1}$.*

Theorem A2.6. *Let $m \geq 3$, and elliptic symbol $\sigma(x,\xi)$ admits n - wave factorization with respect to C_x^{m-n} for $\forall x \in T_n$, $n = 0, 1, \ldots, m-2$. Then operator-function $K(x)$ is homotopic to operator-function $K''(x)$ for which component (A2.8) is an identical operator-function.*

Proof. Indeed, let $\sigma(x,\xi)$ admits the factorization required. The expression

$$\sigma^t(x,\xi) = \exp(t \ln \sigma(x,\xi)), \quad t \in [0;1],$$

is defined correctly, i.e., its logarithm has no bifurcation points because $\sigma(x,\xi)$ nowhere vanishes, and an arbitrary closed curve on $T_n \times S^{m-1}$ is contracted to a point.

Then if

$$\sigma(x,\xi) = \sigma_{\neq}(x,\xi)\sigma_{=}(x,\xi),$$

we have

$$\sigma^t(x,\xi) = \sigma_{\neq}^t(x,\xi)\sigma_{=}^t(x,\xi),$$

and the theorem is proved. ∎

In case $m = 2$ this proof is not suitable in view of following causes. At first we have only corner points ($n = 0$), and second the set $T_0 \times S^1$ is not simply connected. Thus in case $m = 2$ the formulation will be the following.

Theorem A2.6′. *Let $m = 2$, elliptic symbol $\sigma(x,\xi)$ admits 0 - wave factorization with respect to C_x^2, and*

$$\int_{S^1} d \arg \sigma(x,\xi) = 0, \quad \forall x \in T_0. \tag{A2.10}$$

Then the operator-function $K(x)$ is homotopic to operator-function $K'(x)$ for which component (A2.8) is an identical operator-function.

Proof. It will be the same as the proof of Theorem A2.6 taking into account that the condition (A2.10) is quarantee of absence for logarithm of bifurcation points. ∎

Definition A2.2. *Elliptic in point x symbol $\sigma(x,\xi)$ is called non-singular with respect to unit vector $\overline{n}$ if the increment of the argument of function $\sigma(x,\xi)$ for variation from $\sigma(x,-\overline{n})$ to $\sigma(x,\overline{n})$ is contained in interval $(-\pi;\pi)$.*

Lemma A2.2. *Let elliptic symbol $\sigma(x,\xi)$ be given which is non-singular for all unit normals $\overline{n}_x$, $x \in \partial D \setminus \left(\bigcup_{n=0}^{m-2} T_n\right)$.*

For any $\varepsilon > 0$ there is subdomain $D_\varepsilon \subset D$ with smooth boundary ∂D_ε and symbol $\sigma_\varepsilon(x,\xi)$ having the following properties:

1) distance between ∂D and ∂D_ε is not larger than ε;

2) $|\sigma_\varepsilon(x,\xi) - \sigma(x,\xi)| < \varepsilon, \quad \forall x \in \overline{D}, \ \xi \in S^{m-1}$;

3) $\sigma_\varepsilon \in C(\overline{D} \times S^{m-1})$;

4) $\sigma_\varepsilon(x,\xi)$ is non-singular for any normal $\overline{n}_x$, $x \in \partial D_\varepsilon$.

Proof. Let us choose subdomain $D_1 \subset D$ with smooth boundary which is near enough to ∂D, and on normals to ∂D_1, and construct its ε - blowing up D_2 so that $D \subset D_2$. Further we take restriction of $\sigma(x,\xi)$ (on x) on D_1 and continue as a constant on every normal up to ∂D_2. This newly constructed symbol $\sigma_1(x,\xi)$ will be near enough to $\sigma(x,\xi)$ in $\overline{D}$ by virtue of continuity of $\sigma(x,\xi)$. Besides since non-singularity property for $\sigma(x,\xi)$ is stable with respect to small perturbations (it has topological nature) then for $\sigma_1(x,\xi)$ under $x \in \partial D_1$ it may be broken only for those points which are near $\bigcup_{n=0}^{m-2} T_n$.

Let M be a set of such points. In every point x of set $\overline{M}$ (let's recall M is on ∂D_1) we introduce a local coordinate system defined by inner normal $\overline{n}_x$ and tangent plane; thus, coordinates $\overline{n}_x$ are $(0,\ldots,0,1)$.

Let us define function

$$\ell_\delta(x,\xi) = (\xi_m + i|\xi'|)^\delta(\xi_m - i|\xi'|)^{-\delta}$$

in a local coordinate system related to point x, i.e., ξ_m, ξ' are considered in a local coordinate system.

The functions $(\xi_m \pm i|\xi'|)^{\pm\delta}$ are defined as boundary values of functions $(z \pm i|\xi'|)^{\pm\delta}$ (under fixed ξ') that are analytical in the upper and lower complex half-planes (with cut which does not admit bifurcation points). Put

$$\sigma_\delta(x,\xi) = \sigma(x,\xi)\ell_\delta(x,\xi).$$

By a property of the power function we have [128]

$$\frac{\sigma_\delta(x,0,\ldots,0,-1)}{\sigma_\delta(x,0,\ldots,0,+1)} = \exp(2\pi i\delta)\frac{\sigma(x,0,\ldots,0,-1)}{\sigma(x,0,\ldots,0,+1)}.$$

If $\arg\frac{\sigma(x,0,\ldots,0,-1)}{\sigma(x,0,\ldots,0,+1)} = \pi \pmod{2\pi}$ the formula above implies that for small δ $\arg\frac{\sigma_\delta(x,0,\ldots,0,-1)}{\sigma_\delta(x,0,\ldots,0,+1)}$ will be different from π.

(The non-singularity property can be formulated in more convenient terms in the following way; see above. Let $x \in \partial D_\varepsilon$. We construct curve γ_x, $m \geq 3$, or two curves $\gamma_x^\pm$, $m = 2$. In terms of these curves the non-singularity property means that point 0 does not belong to either of these curves, and index of these curves, i.e., degree of corresponding mappings, is equal to zero.

We note that point 0 is not situated on such a curve which is constructed on an approximating symbol. Vanishing of index for such curve follows from the fact that for points outside of $\overline{M}$ the symbol σ_δ is non-singular, the index is integer, and the symbol σ_δ is continuous).

Further we note that under $\delta \to 0$ convergence $\sigma_\delta(x,\xi) \to \sigma(x,\xi)$ is uniform. So, the symbol $\sigma_\delta(x,\xi)$ at point x is non-singular with respect to normal $\overline{n}_x$. If $\overline{m}_x$ is another unit vector which starts from point x and is obtained by rotation ρ of vector $\overline{n}_x$, then we put

$$\sigma_\delta(x,\xi) = \ell_\delta(x,\rho^{-1})\sigma(x,\xi),$$

and thus the required symbol $\sigma_\delta(x,\xi)$ depends indeed on an $(m-1)$ - dimensional parameter Θ which is defined by rotation ρ and varies in an $(m-1)$ - dimensional parallelepiped. Since non-singularity property is stable with respect to small perturbations of vector $\overline{n}_x$ then for every unit vector $\overline{m}_x$ (which can be defined by the same parameter Θ) there is $\delta(\Theta)$ such that $\sigma_\delta(x,\xi)$ is non-singular in some neighbourhood of Θ. Such neighbourhoods obviously form a covering for "parallelepiped of parameters" from which by virtue of compactness of parallelepiped we select a subcovering. So, the function $\delta(\Theta)$ takes no more than a finite number of values. It is left to choose such value of this function that it is less than any other of its non-vanishing values.

Thus, for every point $x \in \overline{M}$ we have constructed the symbol $\sigma_\delta(x,\xi)$ which is non-singular for any unit vector $\overline{m}_x$ which starts from point x. But now δ will depend on x. Taking into account once again the stability of non-singularity property with respect to small variations and applying the method above to $\overline{M}$, we now become free from dependence of δ on x.

Now let us remember that function $\ell_\delta(x,\xi)$ is defined on points $x \in \overline{M}$ only. It is not difficult to achieve that this function will be defined for all $x \in \overline{D_2}$ with preservation for $\sigma_\delta(x,\xi)$ of non-singularity property. Indeed, let us construct subdomain $D_3 \subset D_1$ for which D is ε - blowing up. All points from $D_2 \setminus D_3$ we represent in form (x',t) where x' varies on ∂D_1, and t varies on line segment $[0;2\varepsilon]$. Thus, the points from ∂D_3 are $(x',0)$, the points from ∂D_1 are (x',ε), and the points from ∂D_2 are $(x',2\varepsilon)$. We will construct the function $\ell_\delta(x',\xi)$ with dependence of δ on x' for all $x' \ni \partial D_1 \supset \overline{M}$, and this dependence obviously will be continuous, and then we will introduce the function $\delta_1(x) = \delta(x')t$ and construct $\ell_{\delta_1}(x,\xi)$. It will be defined for all $x \in D_2$ now, and for $x \in D_3$ we have $\ell_{\delta_1}(x',\xi) \equiv 1$.

So, the symbol with properties needed is constructed. ■

Theorem A2.7. *Let elliptic symbol $\sigma(x,\xi)$ be non-singular for $x \in \partial D \setminus \left(\bigcup_{n=0}^{m-2} T_n\right)$. Then operator-function $K(x)$ is homotopic to operator-function $K'(x)$ for which the component (A2.6) in a neighbourhood of ∂D is a multiplication operator on the continuous function $a'(x)$.*

Proof. We construct subdomains D_1, D_2 with smooth boundaries and "overdomain" D_3 so that D_2 is ε - blowing up of D_1, and D_3 is blowing up of

D_2. Further we construct the symbol $\sigma_\delta(x,\xi)$ from Lemma A2.2 so that it is non-singular for $x \in \partial D_3$ (and hence for $x \in \partial D_1$, $x \in \partial D_2$ if ε is sufficiently small). It means that "spectral curves" (see proof of Lemma A2.2), which are constructed on $\sigma_\delta(x,\xi)$, do not intersect the origin, and their indices are equal to 0. As soon as they are such, then they are contractible to a point.

So, "over every point $x \in \partial D_3$" (we mean local coordinate system) $\sigma_\delta(x,\xi)$ is contractible to a point. It follows from Proposition A2.8 (we need an even more simple variant) that homotopy "in leaf" extends to homotopy "in bundle", and thus the symbol $\sigma_\delta(x,\xi)$ is homotopic to $\sigma'_\delta(x)$.

(It is convenient to construct the last homotopy so that parameter t varies on normals between ∂D_1 and ∂D_3 in order to preserve continuous passing from $\sigma_\delta(x,\xi)$ to $\sigma'_\delta(x,\xi)$. ∎

Theorem A2.8. *Let elliptic symbol $\sigma(x,\xi)$ be non-singular for $x \in \partial D \setminus \left(\bigcup_{n=0}^{m-2} T_n\right)$. Then operator-function $K(x)$ is homotopic to operator-function $K''(x)$ for which the component (A2.7) has coefficients not depending on ξ.*

Proof. Essentially we need to repeat the proof of Theorem A2.7. ∎

Before formulating and proving the main result, the index theorem, let us note the following.

Operator A of local type acting in space $L_2(D)$ we will call a virtual Calderon-Zygmund operator if (in Fourier images) it is quasiequivalent to operators

$$\tilde{u}(\xi) \longrightarrow \gamma_m(x,\xi)\tilde{u}(\xi), \quad x \in \mathring{D}, \tag{A2.11}$$

$$\tilde{u}(\xi) \longrightarrow \frac{1+\gamma_{m-1}(x,\xi)}{2}\tilde{u}(\xi) + \frac{1-\gamma_{m-1}(x,\xi)}{2}(H\tilde{u})(\xi),$$
$$x \in \partial D \setminus \left(\bigcup_{n=0}^{m-2} T_n\right), \tag{A2.12}$$

$$\tilde{u}(\xi) \longrightarrow \frac{1+\gamma_n(x,\xi)}{2}\tilde{u}(\xi) - \frac{1-\gamma_n(x,\xi)}{2}(G_{m-n}\tilde{u})(\xi),$$
$$x \in T_n, \quad n = 0,1,\ldots,m-2, \tag{A2.13}$$

where the functions $\gamma_n(x,\xi)$, $n = 0,1,\ldots,m$, satisfy the following conditions: $\gamma_m(x,\xi)$ is defined and continuous in $\overline{D}$, $\gamma_{m-1}(x,\xi)$ is defined and continuous on ∂D, $\gamma_n(x,\xi)$ are defined and continuous on $\overline{T}_n$, $n = 0,1,\ldots,m-2$. On function $\gamma_{m-1}(x,\xi)$ we construct local topological characteristics d_x ($m \geq 3$), $d_x^{\pm}$ ($m = 2$) for $x \in \partial D \setminus \left(\bigcup_{n=0}^{m-2} T_n\right)$ and complete definition for them on the whole boundary ∂D adding all possible limiting states of normals $\overline{n}_y$, $y \to x \in T_n$, $y \in \partial D \setminus \left(\bigcup_{n=0}^{m-2} T_n\right)$.

Then the following holds.

Theorem A2.9. *Let $m \geq 3$. If the following conditions are valid:*

1) $\gamma_m(x,\xi) \neq 0$, $\forall x \in \overline{D}$, $\xi \in S^{m-1}$;
$\gamma_{m-1}(x,\xi) \neq 0$, $\forall x \in \partial D$, $\xi \in S^{m-1}$;
$\gamma_n(x,\xi) \neq 0$, $\forall x \in \overline{T}_n$, $\xi \in S^{m-1}$;

2) $d_x \in (-\pi;\pi)$, $\forall x \in \partial D$, $(m \geq 3)$;
$d_x^{\pm} \in (-\pi;\pi)$, $\forall x \in \partial D$, $(m = 2)$;

3) $\gamma_n(x,\xi)$ *admits* n *- wave factorization with respect to* C_x^{m-n} *for* $\forall x \in \overline{T}_n$ *(and here we add possible limiting states of tangent cone* C_y^{m-n}, $y \to x \in T_{n-1}$, $y \in T_n$*),*

then virtual Calderon-Zygmund operator A *has Noether property in the space* $L_2(D)$.

Proof of this theorem is analogous to proof of Theorem A2.5.

Roughly speaking the operator A is a collection of its local representatives. It is evident that ordinary Calderon-Zygmund operators are virtual at the same time. Actually we will show that for satisfying Noether conditions from Theorem A2.9, the index of any virtual operator is equal to 0, and as a consequence we will obtain a corresponding result on an index of ordinary Calderon-Zygmund operator.

(A virtual Calderon-Zygmund operator is an ordinary Calderon-Zygmund operator only if operators of "higher" and "lower" orders are related one to other in the following way:

$$\gamma_{m-1}(x,\xi) = \gamma_m(x,\Theta_x,\xi),$$

$$\gamma_n(x,\xi) = \gamma_m(x,\varphi_{x,n},\xi)$$

(see above formulas A2.6 — A2.8).

Theorem A2.10. *Let the conditions of Theorem A2.5 (A2.5′) be satisfied. Then the index of operator* K *in the space* $L_2(D)$ *is equal to* 0.

Proof. Let first $m \geq 3$. Taking into account Noetherian conditions for virtual Calderon-Zygmund operators (Theorem A2.9) and by application of Theorems A2.6, A2.7, A2.8, we conclude that it is sufficient to bound oneself by consideration of the following type symbols: $\gamma_m(x,\xi)$ in a neighbourhood of ∂D does not depend on ξ, $\gamma_{m-1}(x,\xi)$ does not depend on ξ, $\gamma_n(x,\xi)$, $n = 0,1,\ldots,m-2$, is identically equal to 1. Such operator-functions (they are invertible) form a group which decomposes into the product of two of its subgroups:

1) $\gamma_m(x,\xi)$ in a neighbourhood of boundary ∂D does not depend on ξ, $\gamma_{m-1}(x,\xi) = \gamma_n(x,\xi) \equiv 1$, $n = 0,1,\ldots,m-2$;

2) $\gamma_m(x,\xi) = \gamma_n(x,\xi) \equiv 1$, $n = 0,1,\ldots,m-2$, $\gamma_{m-1}(x,\xi)$ does not depend on ξ.

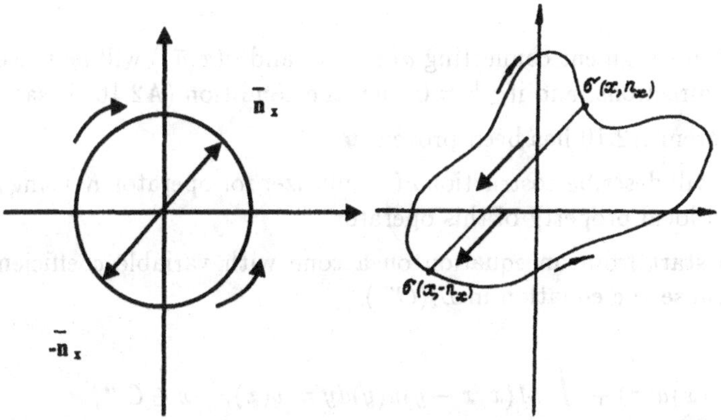

Fig. 1

Each of these subgroups is isomorphic to a group of operators. The first is isomorphic to a group of Noether Calderon-Zygmund operators (up to compact perturbations) in domain D with symbol $\gamma_m(x,\xi)$ not depending on ξ near the boundary, the second is isomorphic to a group of multiplication operators on a non-vanishing continuous function defined on $\partial D \setminus \left(\bigcup_{n=0}^{m-2} T_n\right)$ and having finite limit values on T_n, $n = 0, 1, \dots, m-2$. Thus, every Noether Calderon-Zygmund operator up to small and compact perturbations and homotopies can be "realized" as the product of two operators of type mentioned. Then its index will be equal to sum of indices of these operators. But the index of each of these operators is equal to zero. Index of multiplication operator on non-vanishing functions is equal to zero because it is invertible. For calculation of index of the first operator with symbol not depending on ξ near the boundary, one can use the "pasting of double" operation [225] and reduce the case of manifold with boundary to the case of manifold without boundary with an application of the Atiyah-Singer theorem [13] according to which the index must be zero.

Finally let $m = 2$. In this case we need to apply Theorems **A2.6′**, A2.7, A2.8 and to repeat the arguments. We must explain one point. In Theorem **A2.6′** there is the condition (A2.10), and it is absent in Theorem A2.10. It is found that this condition is satisfied automatically. Let us explain what has been said in more detail (see figure 1).

One the left we draw unit circumference with marked normal ends, and in case $x \in T_0$ under $\overline{n}_x$ one means either left or right normal in point x, and on the right we draw "spectral curve" for which its points are values of symbol $\sigma(x,\xi)$ ($x \in T_0$ is fixed). Let variation of ξ from $-\overline{n}_x$ to $\overline{n}_x$ along the left half-circumference corresponds, for example, to the right piece of "spectral curve" in direction from $\sigma(x,-\overline{n}_x)$ to $\sigma(x,\overline{n}_x)$, and variation along the right half-circumference corresponds to the left piece. We denote these spectral curves $\gamma_x^{\pm}$. Directions are indicated by arrows. It is obvious that

$$\int_{S^1} d \arg \sigma(x,\xi) = \pm(\operatorname{ind} \gamma_x^+ - \operatorname{ind} \gamma_x^-),$$

because the line segment connecting $\sigma(x,-\overline{n}_x)$ and $\sigma(x,\overline{n}_x)$ will be traversed twice in opposite directions, and if $\gamma_x^{\pm} = 0$ then the condition (A2.10) is satisfied also.

So, Theorem A2.10 has been proved. ∎

Now we will describe costruction of regularizer for operator K using a partition of unity and local property of this operator.

We will start from an equation on a cone with variable coefficients of type (A1.19) because the equation in $L_2(C^m)$

$$a(x)u(x) + \int_{C^m} M(x, x-y)u(y)dy = v(x), \quad x \in C^m, \tag{A2.14}$$

is equivalent to equation

$$a_1(x)(P_+U)(x) + \int\limits_{C^m} M_1(x, x-y)U(y)dy + a_2(x)(P_-U)(x) + \\ + \int\limits_{\mathbb{R}^m \setminus C^m} M_2(x, x-y)U(y)dy = V(x), \quad x \in \mathbb{R}^m, \tag{A2.15}$$

in space $L_2(\mathbb{R}^m)$ [128,230] if $a_2(x) \equiv 1$, the kernel M_2 is identically equal to zero, and $V(x)$ coincides with $v(x)$ on C^m and is equal to zero outside of C^m.

The functions $a(x)$, $a_1(x)$, $a_2(x)$ we will assume to be infinite differentiable on $\dot{\mathbb{R}}^m$ (i.e., one point compactification of $\mathbb{R}^m$), and the kernels $M(x, y)$, $M_1(x, y)$, $M_2(x, y)$ are infinite differentiable on $\dot{\mathbb{R}}^m \times (\mathbb{R}^m \setminus \{0\})$, homogeneous of order $-m$ on variable y under fixed x and having vanishing mean value on sphere S^{m-1} in the second variable for all $x \in \dot{\mathbb{R}}^m$.

We will denote the operator

$$u(x) \longrightarrow a(x)u(x) + \int\limits_{\mathbb{R}^m} M(x, x-y)u(y)dy, \quad x \in \mathbb{R}^m,$$

by letter M, and then the equation (A2.14) in operator form will be

$$P_+ M u_+ = v, \tag{A2.16}$$

(the index "+" of function u indicates that it is given in C^m); analogously, the operators

$$U(x) \longrightarrow a_i(x)U(x) + \int\limits_{\mathbb{R}^m} M_i(x, x-y)U(y)dy, \quad x \in \mathbb{R}^m, \quad i = 1, 2,$$

will be denoted by M_1 and M_2, and the equation (A2.15) itself we will write in form

$$(M_1 P_+ + M_2 P_-)U = V. \tag{A2.17}$$

Let us distinguish the relation between inverse operators for $P_+M : L_2(C^m) \rightarrow L_2(C^m)$ and $MP_+ + P_- : L_2(\mathbb{R}^m) \rightarrow L_2(\mathbb{R}^m)$: they are invertible or not at the same time [128], and if operator $MP_+ + P_-$ is invertible then the inverse operator for P_+M is an operator $P_+(MP_+ + P_-)^{-1}$. Taking into account this relation we will study the equation (A2.17).

Let us fix point x_0 and consider the operator

$$M_{i,x_0} : U(x) \longrightarrow a_i(x_0)U(x) + \int\limits_{\mathbb{R}^m} M_i(x, x-y)U(y)dy, \\ x \in \mathbb{R}^m, \quad i = 1, 2,$$

and the equation

$$(M_{1,x_0} P_+ + M_{2,x_0} P_-)U = V. \tag{A2.18}$$

The equation (A2.18) is the equation (A1.19) explicitly as soon as the point x_0 is fixed, and it is studied by corresponding methods (see Appendix 1).

Let us denote symbols of operators M_i, $i = 1, 2$, by $\sigma_i(x,\xi)$ and operators with symbols

$$\frac{\sigma_2(x,\xi)+\sigma_1(x,\xi)}{2\sigma_1(x,\xi)\sigma_2(x,\xi)}, \quad \frac{\sigma_2(x,\xi)-\sigma_1(x,\xi)}{2\sigma_1(x,\xi)\sigma_2(x,\xi)}$$

we denote A and B respectively.

Definition A2.3. *Let $\sigma(x,\xi)$ be an elliptic symbol from class $C^\infty\left(\dot{\mathbb{R}}^m \times S^{m-1}\right)$. We say that $\sigma(x,\xi)$ admits smooth n - wave factorization with respect to C^m if for fixed x we have factorization in the sense of Definition A1.1*

$$\sigma(x,\xi) = \sigma_{\neq}(x,\xi)\sigma_{=}(x,\xi,)$$

and the elements of factorization are differentiable on x.

Theorem A2.11. *Let $\sigma_1^{-1}(x,\xi)\sigma_2(x,\xi) \in C^\infty\left(\dot{\mathbb{R}}^m \times S^{m-1}\right)$, $\sigma_1^{-1}(x,\xi)\sigma_2(x,\xi) \neq 0$, $\forall x \in \dot{\mathbb{R}}^m$, $\xi \in S^{m-1}$. If $\sigma_1^{-1}(x,\xi)\sigma_2(x,\xi)$ admits smooth 0 - wave factorization with respect to C^m for $\forall x \in \dot{\mathbb{R}}^m$, then operator*

$$M_1P_+ + M_2P_-$$

has a two-side regularizer in space $L_2(\mathbb{R}^m)$ of type

$$A - BN_{\neq}P_+N_{\neq}^{-1},$$

where $N_{\neq}^{\pm 1}$ is a multidimensional singular integral operator with symbol $\pm W_{\neq}^{\pm 1}(x,\xi)\sigma_1^{\pm}(x,\xi)$.

Proof. Let us show that

$$\left(A - BN_{\neq}P_+N_{\neq}^{-1}\right)(M_1P_+ + M_2P_-) = I + T, \qquad (A2.19)$$

where T is a compact operator in space $L_2(\mathbb{R}^m)$.

We denote kernels of operators A, B and so on, by the same letter with corresponding arguments.

At once we will note (it is the key point of our proof) that if for the kernels A, B, $N_{\neq}^{\pm 1}$ we fix the pole x_0 and consider the operators A_{x_0}, B_{x_0}, $N_{\neq,x_0}^{\pm 1}$ respectively, then obviously

$$\left(A_{x_0} - B_{x_0}N_{\neq,x_0}P_+N_{\neq,x_0}^{-1}\right)(M_{1,x_0}P_+ + M_{2,x_0}P_-) = I. \qquad (A2.20)$$

Let us introduce into consideration "kernels" $M(x,y)$ and $Z(x,y)$ of operators $M_1P_+ + M_2P_-$ and $N_{\neq}P_+$ in the following way:

$$\int_{\mathbb{R}^m} M(x, x-y)u(y)dy \stackrel{\text{def}}{\equiv}$$

$$\stackrel{\text{def}}{\equiv} \int\limits_{C^m} M_1(x, x-y)u(y)dy + \int\limits_{\mathbb{R}^m \setminus C^m} M_2(x, x-y)u(y)dy,$$

$$\int\limits_{\mathbb{R}^m} Z(x, x-y)u(y)dy \stackrel{\text{def}}{\equiv} \int\limits_{C^m} N_{\neq}(x, x-y)u(y)dy.$$

Now the left-hand part of formula (A2.19) will be written in form

$$A(x, x-y) \circ M(y, y-\eta) -$$

$$-B(x, x-y) \circ Z(y, y-t) \circ N_{\neq}^{-1}(t, t-h) \circ M(h, h-\eta)$$

(arguments are introduced to avoid confusion), where sign "o" indicates successive action of operators with corresponding kernels.

The first summand can be represented as

$$A(x, x-y) \circ M(y, y-\eta) = A(x, x-y) \circ M(x, y-\eta) +$$

$$+A(x, x-y) \circ [M(y, y-\eta) - M(x, y-\eta)].$$

Analogously we write

$$N_{\neq}^{-1}(t, t-h) \circ M(h, h-\eta) = N_{\neq}^{-1}(t, t-h) \circ M(t, h-\eta) +$$

$$+N_{\neq}^{-1}(t, t-h) \circ [M(h, h-\eta) - M(t, h-\eta)]$$

and denote

$$N_{\neq}^{-1}(t, t-h) \circ M(t, h-\eta) \stackrel{\text{def}}{\equiv} L(t, t-\eta)$$

the anew obtained "kernel" of Calderon-Zygmund.

It is admissible by the following reasons:

$$\int\limits_{\mathbb{R}^m} N_{\neq}^{-1}(t, t-h) \left(\int\limits_{C^m} M_1(t, h-\eta)u(\eta)d\eta + \int\limits_{\mathbb{R}^m \setminus C^m} M_2(t, h-\eta)u(\eta)d\eta \right) =$$

$$= \int\limits_{\mathbb{R}^m} N_{\neq}^{-1}(t, t-h) \left(\int\limits_{\mathbb{R}^m} M_1(t, h-\eta)U(\eta)d\eta + \int\limits_{\mathbb{R}^m} M_2(t, h-\eta)V(\eta)d\eta \right)$$

where

$$U(\eta) = \begin{cases} u(\eta), \ \eta \in C^m \\ 0, \ \eta \notin C^m \end{cases}, \qquad V(\eta) = \begin{cases} 0 \ \eta \in C^m \\ u(\eta), \ \eta \in \mathbb{R}^m \setminus C^m \end{cases}.$$

All kernels in the last representation are real kernels of Calderon-Zygmund, and with fixed pole. It is well-known that such kernels like ordinary convolutions [153], particularly convolution of two such kernels, meaning in principal value sense [153], generates anew kernels of Calderon-Zygmund with possible addition $a(t)u(t)$ depending on whether or not the new kernel has vanishing mean value on

S^{m-1}. This addition (as is easily seen) does not affect the course of the following arguments, and so we will mean that this addition is absent.

Denoting

$$N_{\neq}^{-1}(t,t-h)\circ M_1(t,h-\eta)\equiv L_1(t,t-\eta),$$

$$N_{\neq}^{-1}(t,t-h)\circ M_2(t,h-\eta)\equiv L_2(t,t-\eta)$$

the operator with "kernel" $L(t,t-\eta)$ we will mean

$$\int_{\mathbb{R}^m} L(t,t-\eta)u(\eta)d\eta =$$

$$= \int_{C^m} L_1(t,t-\eta)u(\eta)d\eta + \int_{\mathbb{R}^m\setminus C^m} L_2(t,t-\eta)u(\eta)d\eta.$$

Note that "kernel" $L(t,t-\eta)$ preserves smoothness properties on the first variable of kernels $N_{\neq}^{-1}$, M_i, $i=1,2$.

After similar transformations of "kernel" Z we obtain in the last analysis the following:

$$A(x,x-y)\circ M(y,y-\eta)-$$

$$-B(x,x-y)\circ Z(y,y-t)\circ N_{\neq}^{-1}(t,t-h)\circ M(h,h-\eta)=$$

$$= A(x,x-y)\circ M(y,y-\eta)-$$

$$-B(x,x-y)\circ Z(x,y-t)\circ N_{\neq}^{-1}(x,t-h)\circ M(x,h-\eta)+$$

+ summands each of which has operator of one of two types as a factor:

$$1)\quad \int_{\mathbb{R}^m} A_1(x,x-y)\left(\int_{C^m}\left[A_2(y,y-t)-A_2(x,y-t)\right]u(t)dt\right)dy,$$

$$2)\quad \int_{\mathbb{R}^m} A_1(x,x-y)\left(\int_{\mathbb{R}^m\setminus C^m}\left[A_2(y,y-t)-A_2(x,y-t)\right]u(t)dt\right)dy, \tag{A2.21}$$

where $A_i(x,y)$, $i=1,2$, are kernels of Calderon-Zygmund.

The operators (A2.21) are roughly speaking the operators with weak singularity (i.e., potential type operators). Indeed let us decompose the kernels $A_2(x,y)$ on sphere S^{m-1} into series of spherical harmonics in variable y (smoothness conditions, which are assumed on kernels, permit us to do so [171,172]). Then

$$\int_{C^m}\left[A_2(y,y-t)-A_2(x,y-t)\right]u(t)dt =$$

$$= \int_{C^m}\sum_{n=0}^{\infty}\sum_{k=1}^{k_n}\left[a_n^{(k)}(x)-a_n^{(k)}(y)\right]Y_{n,m}^{(k)}(\Theta)|y-t|^{-m}u(t)dt,$$

where $\Theta = \frac{y-t}{|y-t|}$, and the series uniformly converges on S^{m-1}.

Thus,

$$\int\limits_{\mathbb{R}^m} A_1(x, x-y) \left(\int\limits_{C^m} [A_2(y, y-t) - A_2(x, y-t)]\, u(t)dt \right) dy =$$

$$= \sum_{n=0}^{\infty} \sum_{k=1}^{k_n} \int\limits_{\mathbb{R}^m} A_1(x, x-y) \left[a_n^{(k)}(x) - a_n^{(k)}(y) \right] \left(\int\limits_{C^m} A_{n,m}^{(k)}(y-t)u(t)dt \right) dy,$$

where $A_{n,m}^{(k)}(y-t) = Y_{n,m}^{(k)}(\Theta)|y-t|^{-m}$ is a kernel of Calderon-Zygmund.

Interior integrals will be defined as bounded operators $L_2(C^m) \to L_2(\mathbb{R}^m)$, and exterior ones will be defined as compact operators because $|a_n^{(k)}(x) - a_n^{(k)}(y)| \leq c_{n,k}|x-y|$, and hence exterior integral has weak singularity only. Since an operator series converges in an operator topology then as a result we will obtain a compact operator. Let us note that similar arguments were used by S.G.Mikhlin under proof of symbol multiplication rule [171]. So,

$$A(x, x-y) \circ M(y, y-\eta) -$$

$$- B(x, x-y) \circ Z(y, y-t) \circ N_{\neq}^{-1}(t, t-\eta) \circ M(h, h-\eta) = I + T$$

(summand I has been obtained by virtue of formula (A2.20)).

We have proved that operator $A - BN_{\neq}P_+N_{\neq}^{-1}$ is a left regularizer for operator $M_1P_+ + M_2P_-$ in space $L_2(\mathbb{R}^m)$. Analogously, one can verify it is a right regularizer. ∎

Corollary A2.1. *Let M be a Calderon-Zygmund operator with symbol $\sigma(x,\xi)$, $\sigma(x,\xi) \in C^\infty\left(\overline{C^m} \times S^{m-1}\right)$, $\sigma(x,\xi) \neq 0$, $\forall x \in \overline{C^m}$, $\xi \in S^{m-1}$. Let us denote $\tilde{A}$, $\tilde{B}$ as Calderon-Zygmund operators with symbols*

$$\frac{1+\sigma(x,\xi)}{2\sigma(x,\xi)}, \quad \frac{1-\sigma(x,\xi)}{2\sigma(x,\xi)}.$$

If $\sigma(x,\xi)$ admits smooth 0 - wave factorization with respect to C^m for $\forall x \in \overline{C^m}$

$$\sigma(x,\xi) = \sigma_{\neq}(x,\xi)\sigma_{=}(x,\xi),$$

then operator P_+M : $L_2(C^m) \to L_2(C^m)$ has a two-side regularizer in space $L_2(C^m)$ which is represented in form

$$P_+\left(\tilde{A} - \tilde{B}\tilde{N}_{\neq}P_+\tilde{N}_{\neq}^{-1}\right),$$

where $\tilde{N}_{\neq}^{\pm 1}$ is a Calderon-Zygmund operator with symbol $\pm\sigma_{=}^{\pm 1}(x,\xi)$.

The case of wedge $\mathbb{R}^n \times C^{m-n}$ can be considered analogously with the help of Theorem A1.5 (see Appendix 1).

Further our goal is to construct a regularizer for operator K in bounded domain D. This regularizer will be collected "by pieces" taking into account specific character of local representatives in different points of domain D.

Let domain D be covered by a finite number of domains $\{D_j\}_{j=1}^r$, and $\{\varphi_j(x)\}_{j=1}^r$ be a partition of identity subordinated to covering $\{D_j\}_{j=1}^r$, $\sum_{j=1}^r \varphi_j(x) = 1$, $x \in \overline{D}$, and let functions $\psi_j(x)$, $j = 1, 2, \ldots, r$, be such that $\psi_j(x) \in C_0^\infty(\mathbb{R}^m)$, $\psi_j(x) = 1$ on $\operatorname{supp} \varphi_j$, and supports of $\varphi_j(x)$ and $1 - \psi_j(x)$ do not intersect.

Without loss of generality one can mean that the covering is chosen in the following way. At first the points $x_j \in T_n$ and balls D_j with center in these points are taken so that these balls cover some neighbourhood of T_n, $n = 0, 1, \ldots, m-2$. Then the points $x_j \in \partial D \setminus \left(\bigcup_{n=0}^{m-2} T_n \right)$ and balls D_j are taken so that these balls in addition to the first ones cover some neighbourhood of ∂D. And then points $x_j \in \overset{\circ}{D}$ are taken so that balls D_j with center in these points in addition to the balls taken earlier give a covering of the whole domain D.

Let us denote R_+ the projector on domain D :

$$(R_+ u)(x) = \begin{cases} u(x), & x \in D \\ 0, & x \notin \overline{D}, \end{cases}$$

and write operator K in form

$$R_+ K u_+,$$

where function u_+ is defined in domain D.

Further let us note the following:

$$R_+ K u_+ = \sum_{j=1}^r \varphi_j R_+ K u_+ = \sum_{j=1}^r R_+ \varphi_j K u_+ =$$

$$= \sum_{j=1}^r R_+ \varphi_j K \psi_j u_+ + \sum_{j=1}^r R_+ \varphi_j K (1 - \psi_j) u_+ =$$

$$= \sum_{j=1}^r R_+ \varphi_j K \psi_j u_+ + T u_+,$$

where T is a compact operator because supports φ_j and $1 - \psi_j$ do not intersect, and thus the kernel has no singularity.

Let us consider different cases of location for elements of covering D_j. Let $\sigma(x, \xi)$ be symbol of operator K; we assume $\sigma(x, \xi) \neq 0$, $\forall x \in \overline{D}$, $\xi \in S^{m-1}$.

1) $D_j \cap \partial D = \emptyset$, i.e., neighbourhood D_j is interior. We construct the kernel $K_j(x, y)$ on symbol $\sigma^{-1}(x, \xi)$ for $x \in D_j$ and corresponding singular operator

$$(K_j'' u)(x) = \text{ v.p.} \int_{D_j} K_j(x, x - y) u(y) dy, \quad x \in D_j,$$

putting

$$K_j \equiv \psi_j K_j'' \varphi_j.$$

2) $D_j \cap \partial D \neq \emptyset$, $D_j \cap T_n = \emptyset$, $n = 0, 1, \ldots, m-2$, i.e., D_j "serves" a smooth piece of boundary ∂D. Let x_j be center of ball $\mathfrak{D}_j$. Let us introduce a local coordinate system related to point x_j taking x_j as origin, directing m-th axis on inner normal to ∂D in point x_j, and left $m-1$ axes we arrange in a tangent plane. We denote local coordinates by (ξ, η) and equation $\partial D \cap D_j$ we write as $\eta = F_{x_j}(\xi)$. It is evident that according to our assumptions the function $F_{x_j}(\xi)$ is infinitely differentiable.

Let us consider mapping $\tilde{\varphi}_j$ of domain D_j which is defined in the following way: if $y = (\xi, \eta)$ then $\tilde{\varphi}_j(y) \equiv \tilde{y} = (\xi, \eta - F_{x_j}(\xi))$. Obviously under this mapping the points x_j remains in its place, $\partial D \cap D_j$ is transformed into a piece of a hyperplane which is a tangent plane to ∂D in point x_j, and Jacobian of mapping $\tilde{\varphi}_j$ has the form

$$\begin{vmatrix} 1 & 0 & \ldots & \ldots & 0 \\ 0 & 1 & \ldots & \ldots & 0 \\ \ldots & \ldots & \ldots & \ldots & \ldots \\ 0 & 0 & \ldots & 1 & 0 \\ -\frac{\partial F_{x_j}}{\partial \xi_1} & -\frac{\partial F_{x_j}}{\partial \xi_2} & \ldots & -\frac{\partial F_{x_j}}{\partial \xi_{m-1}} & 1 \end{vmatrix},$$

and so is equal to 1.

Let us denote $\tilde{\varphi}_j(U_1) = V_1$, where U_1 is support of function $\psi_j(x)$, $t = \tilde{\varphi}_j(y)$, $\tau = \tilde{\varphi}_j(x)$. Then (see above also)

$$K(x, x-y) = K\left(\tilde{\varphi}_j^{-1}(\tau), (\tilde{\varphi}_j^{-1})'(\tau)(\tau - t)\right) + R(\tau, t), \quad x, y \in U_1,$$

where the kernel $R(\tau, t)$ has weak singularity, and hence it generates an integral operator of potential type which is compact.

Let us introduce other notations: we will denote by $E(\tau)$ the matrix $(\tilde{\varphi}^{-1})'(\tau)$ and put

$$K((\tilde{\varphi}_j^{-1}(\tau), E(\tau)(\tau - t)) \equiv \tilde{K}(\tau, \tau - t), \quad \tau \in V_1.$$

The classical symbol of Calderon-Zygmund operator with kernel $\tilde{K}(\tau, t)$ will be denoted by $\tilde{\sigma}(\tau, \xi)$ It is well-known [171] that

$$\tilde{\sigma}(\tau, \xi) = \sigma\left(\tilde{\varphi}_j^{-1}(\tau), (E^*)(\tau)\xi\right),$$

where $E^*(\tau)$ is the matrix conjugate to matrix $E(\tau)$.

Thus in our case

$$R_+\varphi_j K\psi_j u_+ = R'_{+,x_j}\overline{\varphi}_j\tilde{K}\overline{\psi}_j U_+ + R'_{+,x_j}\overline{\varphi}_j R\overline{\psi}_j U_+, \tag{A2.22}$$

where R_{+,x_j} denotes a projector on $\mathbb{R}^m_+$ in a local coordinate system related to point x_j,

$$\overline{\varphi}_j \equiv \varphi_j \circ \tilde{\varphi}_j^{-1}, \quad \overline{\psi}_j \equiv \psi_j \circ \tilde{\varphi}_j^{-1}, \quad U_+ \equiv u_+ \circ \tilde{\varphi}_j^{-1}.$$

Now we will use some results from paper [4] in which the inverse operator is constructed in half-space for an operator with kernel $\tilde{K}(t)$ not depending on pole τ. The results of paper [4] are related to weighted Hölder spaces, and therefore we'll insert corresponding corrections for space L_2. (What we speak of below is related to the theory of one-dimensional singular integral equations; details can be found in [108,128]).

Let us consider the equation in space $L_2(\mathbb{R}^m)$

$$\left(\tilde{K}_1\Pi_+ + \tilde{K}_2\Pi_-\right)u = v, \tag{A2.23}$$

where $\Pi_\pm$ is a projector on $\mathbb{R}^m_\pm$, $\tilde{K}_1$, $\tilde{K}_2$ are Calderon-Zygmund operators with kernels $\tilde{K}_1(\tau,t)$, $\tilde{K}_2(\tau,t)$ and symbols $\tilde{\sigma}_1(\tau,\xi)$, $\tilde{\sigma}_2(\tau,\xi)$ respectively.

We fix the pole τ for kernels $\tilde{K}_1(\tau,t)$, $\tilde{K}_2(\tau,t)$ and consider the equation (A2.23) as an equation with constant coefficients. By Fourier transform it reduces to equation

$$[\tilde{\sigma}_1(\tau,\xi',\cdot)P_{\xi'} + \tilde{\sigma}_2(\tau,\xi',\cdot)Q_{\xi'}]\,\tilde{u} = \tilde{v}$$

with parameter ξ'; here $\tilde{u}$, $\tilde{v}$ denote Fourier transforms of functions u, v,

$$P_{\xi'} = \frac{1}{2}(I - H), \quad Q_{\xi'} = \frac{1}{2}(I + H).$$

Through straight line ξ_m and point ξ' we construct complex plane $\mathbb{C}_{\xi'}$ and define on it a pair of functions ($m > 2$)

$$\omega_\pm(\tau,\xi',z) = (z \pm i|\xi'|)^{-\gamma(\tau)},$$

where

$$\gamma(\tau) = \frac{1}{2\pi i}\ln\frac{\left(\tilde{\sigma}_1\tilde{\sigma}_2^{-1}\right)(\tau,0,\ldots,0,-1)}{\left(\tilde{\sigma}_1\tilde{\sigma}_2^{-1}\right)(\tau,0,\ldots,0,+1)},$$

and the value

$$\arg\frac{\left(\tilde{\sigma}_1\tilde{\sigma}_2^{-1}\right)(\tau,0,\ldots,0,-1)}{\left(\tilde{\sigma}_1\tilde{\sigma}_2^{-1}\right)(\tau,0,\ldots,0,+1)}$$

we subordinate to restriction

$$-\frac{1}{2} < \operatorname{Re}\gamma(\tau) < \frac{1}{2}; \tag{A2.24}$$

in plane $\mathbb{C}_{\xi'}$ we make the cut which forbids the going around infinite point.

(In case $m = 2$ one needs to consider the functions $\omega_\pm(\tau,\xi',z) = (z \pm i\xi')^{-\gamma(\tau)}$).

Let us denote limit values of functions $\omega_\pm$ on straight line ξ_m by

$$\omega(\tau,\xi',\xi_m) = (\xi_m \pm i|\xi'|)^{-\gamma(\tau)}$$

and put

$$\omega(\tau,\xi',\xi_m) = \left(\frac{\xi_m - i|\xi'|}{\xi_m + i|\xi'|}\right)^{\gamma(\tau)}.$$

On function $\left(\tilde{\sigma}_1\tilde{\sigma}_2^{-1}\omega\right)(\tau,\xi',\xi_m)$ we define topological characteristic d_τ (see above) for $m > 2$ and $d_\tau^\pm$ for $m = 2$ and require of their vanishing:

$$d_\tau = 0, \quad m > 2; \quad d_\tau^\pm = 0, \quad m = 2. \tag{A2.25}$$

Further we define two functions which are homogeneous of order zero:

$$g_{\pm}(\tau,\xi',\xi_m)=\exp\left(\frac{1}{2}\ln\left(\tilde{\sigma}_1\tilde{\sigma}_2^{-1}\omega^{-1}\right)(\tau,\xi',\xi_m)\mp\right.$$

$$\left.\mp\frac{1}{2\pi i}\ \text{v.p.}\int\limits_{-\infty}^{+\infty}\frac{\ln\left(\tilde{\sigma}_1\tilde{\sigma}_2^{-1}\omega^{-1}\right)(\tau,\xi',\eta)d\eta}{\xi_m-\eta}\right)$$

and construct the operator

$$\begin{aligned}\left(\tilde{K}_1\Pi_+ + \tilde{K}_2\Pi_-\right)^{-1} &\equiv \left[g_+^{-1}(\tau,\xi',\cdot)P_{\xi'}+g_-^{-1}(\tau,\xi',\cdot)Q_{\xi'}\right]\times\\ \times&\left[\frac{\omega^{-1}(\tau,\xi',\cdot)+1}{2}I+\frac{\omega^{-1}(\tau,\xi',\cdot)-1}{2}\left(\omega_-^{-1}(\tau,\xi',\cdot)H\omega_-(\tau,\xi',\cdot)\right)\right]\times\\ \times& g_-^{-1}(\tau,\xi',\cdot)\tilde{\sigma}_2^{-1}(\tau,\xi',\cdot)\end{aligned} \tag{A2.26}$$

The operator (A2.26) (it is written in Fourier images) will be regularizer for operator $\tilde{K}_1\Pi_+ + \tilde{K}_2\Pi_-$. In order to verify it one needs to apply inverse Fourier transform (as a result the operator (A2.26) is transformed into sum of products of Calderon-Zygmund operators and operators $\Pi_\pm$) and essentially to repeat the proof of the Theorem A2.11.

Let us recall now what we really need in a regularizer for operator $\Pi_+\tilde{K}\ :\ L_2(\mathbb{R}^m_+)\to L_2(\mathbb{R}^m_+)$, but knowing a regularizer for operator $\tilde{K}\Pi_+ + I\Pi_-\ :\ L_2(\mathbb{R}^m)\to L_2(\mathbb{R}^m)$ we find that a regularizer for operator $\Pi_+\tilde{K}$ is the operator $\Pi_+\left(\tilde{K}\Pi_+ + I\Pi_-\right)^{-1}$.

So, in case 2) by satisfying conditions (A2.24), (A2.25), $\tau\in V_1$ we construct the operator

$$\tilde{K}_j'' = R'_{+,x_j}\left(\tilde{K}R_{+,x_j}+IR'_{-,x_j}\right)^{-1},$$

the operator

$$K_j' = \overline{\psi}_j\tilde{K}_j''\overline{\varphi}_j, \tag{A2.27}$$

and operator K_j obtained from (A2.27) by inverse change of variables (under this operation, generally speaking, compact addition can appear),

$$K_j = \psi_j K_j'\varphi_j.$$

3) $D_j\cap\partial D\neq\emptyset$, $k=\min\{n\}$, $D_j\cap T_n\neq\emptyset$, $n=0,1,\ldots,m-2$.

To avoid repetition, we will describe only the modifications one needs to insert in the arguments above to obtain a local regularizer in case 3).

Existence of diffeomorphisms needed permits us to apply an analogous change of variables and to reduce the question to model equation (A2.17). Further one applies Corollary A2.1 and fully repeats the arguments from 2). Under this approach one additional condition appears: for $x\in D_j$ symbol $\sigma(x,\xi)$ admits smooth k – wave factorization with respect to $C^{m-k}_{x_j}$.

Now collecting these local regularizers, constructed according to the specific character of covering elements D_j, one can declare that operator

$$K^{-1} = \sum_{j=1}^{r} K_j$$

is regularizer for operator K.

Let us verify it.

$$K^{-1}(R_+K)u_+ = \sum_{j=1}^{r} \psi_j K_j'' \varphi_j R_+ K u_+ =$$

$$= \sum_{j=1}^{r} \psi_j K_j'' \varphi_j R_+ K \psi_j u_+ + \sum_{j=1}^{r} \psi_j K_j'' \varphi_j R_+ K(1-\psi_j) u_+ =$$

$$= \sum_{j=1}^{r} \psi_j K_j'' \varphi_j R_+ K \psi_j u_+ + T u_+ =$$

(here local change of variables above for cases 2), 3) is applied; in case 1) one does not need a change of variables, but we will not distinguish specially these neighbourhoods)

$$= \sum_{j=1}^{r} \overline{\psi}_j \tilde{K}_j'' \overline{\varphi}_j \tilde{K}_j \overline{\psi}_j U_+ + \tilde{T} U_+ =$$

(because singular operators and multiplication operators on smooth functions commute up to compact operators [171])

$$= \sum_{j=1}^{r} \overline{\varphi}_j \tilde{K}_j'' \tilde{K}_j \overline{\psi}_j U_+ + \tilde{T}_1 U_+ =$$

$$= \sum_{j=1}^{r} \overline{\varphi}_j U_+ + \tilde{T}_2 U_+ = \sum_{j=1}^{r} \varphi_j u_+ + T_3 u_+ =$$

$$= u_+ + T_3 u_+ = (I + T_3) u_+.$$

So, K^{-1} is left regularizer for operator K, and analogously one can verify that it is right regularizer also. ■

Remark A2.2. Conditions (A2.24) and (A2.25) are stable with respect to small perturbations, so that choosing enough "fine" covering one can require satisfying of these conditions in boundary points only. In the case of smooth k - wave factorization, explicit construction of a regularizer requires its existence in any small neighbourhood of submanifolds T_n, $n = 0, 1, \ldots, m-2$.

Appendix 3: The Mellin transform

The Mellin transform is defined by formula

$$\hat{f}(s) = \int_0^\infty f(x)x^{s-1}dx, \quad s = \sigma + i\tau, \tag{A3.1}$$

at least for functions $f(x) \in C_0^\infty(\mathbb{R}_+)$. Integral (A3.1) converges for all complex s and it is an entire analytic function. If we change variable $x = e^t$ then the Mellin transform passes into the Fourier transform of function $f(e^t)$:

$$\hat{f}(s) = \int_{-\infty}^\infty e^{(\sigma+i\tau)} f(e^t)dt, \quad s = \sigma + i\tau. \tag{A3.2}$$

Thus, all properties of the Mellin transform can be obtained from corresponding properties of the Fourier transform. Particularly, the inversion formula of the Mellin transform for $f(x) \in C_0^\infty(\mathbb{R})$ has the following form:

$$f(x) = \frac{1}{2\pi}\int_{-\infty}^\infty \hat{f}(s)t^{-s}d\tau, \quad s = \sigma + i\tau. \tag{A3.3}$$

Parceval equality for Mellin transform:

$$\int_0^{+\infty} t^{2\sigma-1}|f(t)|^2dt = \frac{1}{2\pi}\int_{-\infty}^{+\infty} \left|\hat{f}(s)\right|^2 d\tau, \quad s = \sigma + i\tau, \tag{A3.4}$$

particularly, for $\sigma = 1/2$ we have

$$\int_0^{+\infty} |f(t)|^2dt = \frac{1}{2\pi}\int_{-\infty}^{+\infty} \left|\hat{f}(s)\right|^2 d\tau, \quad s = 1/2 + i\tau,$$

or, in other words,

$$\int_0^{+\infty} |f(t)|^2dt = \frac{1}{2\pi i}\int_{1/2-i\infty}^{1/2+i\infty} \left|\hat{f}(s)\right|^2 ds, \tag{A3.5}$$

meaning the right integral as

$$\lim_{y\to\infty} \int_{1/2-iy}^{1/2+iy} \left|\hat{f}(s)\right|^2 ds.$$

Let us describe basic properties of the Mellin transform. We will start from transformations of operations, and for our convenience we will denote the Mellin transform by letter M also.

1) $M\left[(\ln x)^k f(x)\right] = \dfrac{d^k}{ds^k}\hat{f}(s);$

2) $M\left[\dfrac{d^k}{dx^k}f(x)\right] = (-1)^k(s-1)(s-2)\cdots(s-k)\hat{f}(s-k);$

3) $M\left[\dfrac{d^k}{dx^k}\left(x^k f(x)\right)\right] = (-1)^k(s-1)(s-2)\cdots(s-k)\hat{f}(s);$

4) $M\left[x^\alpha f(x)\right] = \hat{f}(s+\alpha);$

5) $M\left[\left(\dfrac{d}{dx}x\right)^k f(x)\right] = (-1)^k(s-1)\hat{f}(s);$

6) $M\left[f\left(x^r\right)\right] = |r|^{-1}\hat{f}\left(r^{-1}s\right), \quad r \neq 0;$

7) $M\left[f(rx)\right] = r^{-s}\hat{f}(s), \quad r > 0.$

Proposition A3.1. *Let $x^{s-1}f(x) \in L(0;+\infty)$, and the function $f(x)$ have bounded variation in a neighbourhood of point $x = t$. Then*

$$\frac{f(t+0)+f(t-0)}{2} = \frac{1}{2\pi i}\lim_{y\to\infty}\int_{\sigma-iy}^{\sigma+iy}\hat{f}(s)t^{-s}ds,$$

where the function $\hat{f}(s)$ is defined by formula (A3.1).

Proposition A3.2. *Let $\hat{f}(\sigma+i\tau) \in L(-\infty;+\infty)$ and have bounded variation in a neighbourhood of point $\tau = t$. Then*

$$\frac{1}{2}\left[\hat{f}(\sigma+i((\tau+0))+\hat{f}(\sigma+i(\tau-0))\right] = \lim_{y\to\infty}\int_{1/y}^{y} f(x)x^{\sigma+ix-1}dt,$$

where $f(x) = \frac{1}{2\pi i}\int\limits_{\sigma-i\infty}^{\sigma+i\infty}\hat{f}(s)x^{-s}ds.$

Let now $\hat{f}(s)$ and $\hat{g}(s)$ be Mellin transforms of functions $f(x)$ and $g(x)$. Immediately one can obtain

$$\begin{aligned}\frac{1}{2\pi i}\int_{k-i\infty}^{k+i\infty}\hat{f}(s)\hat{g}(1-s)ds &= \frac{1}{2\pi i}\int_{k-i\infty}^{k+i\infty}\hat{g}(1-s)\int_0^{+\infty}f(x)x^{s-1}dx = \\ &= \frac{1}{2\pi i}\int_0^{+\infty}f(x)dx\int_{k-i\infty}^{k+i\infty}\hat{g}(1-s)x^{s-1}ds = \int_0^{+\infty}f(x)d(x)dx.\end{aligned} \tag{A3.6}$$

In a similar way we have

$$\frac{1}{2\pi i}\int\limits_{k-i\infty}^{k+i\infty}\hat{f}(s)\hat{g}(s)ds=\frac{1}{2\pi i}\int\limits_{0}^{+\infty}g(x)dx\int\limits_{k-i\infty}^{k+i\infty}\hat{f}(s)x^{s-1}ds=$$
$$=\int\limits_{0}^{+\infty}g(x)f\left(\frac{1}{x}\right)\frac{dx}{x}. \qquad (A3.7)$$

Justification of this result is given by the following

Proposition A3.3. *Let either*

$$x^{k-1}f(x)\in L(0;+\infty) \text{ and } \hat{g}(1-k-ix)\in L(0;+\infty)$$

or

$$\hat{f}(k+ix)\in L(-\infty;+\infty) \text{ and } x^k g(x)\in L(0;+\infty).$$

Then the formula (A3.6) is valid.

More important for us is the following analogue of the convolution theorem.

Proposition A3.4. *Let $x^k f(x)$, $x^k g(x)\in L(0;+\infty)$, and*

$$h(y)=\int\limits_{0}^{+\infty}f(x)g\left(\frac{y}{x}\right)\frac{dx}{x}.$$

Then the function $y^k h(y)$ belongs to $L(0;+\infty)$, and its Mellin transform is $\hat{f}(s)\hat{g}(s)$.

We will give some additional examples of Mellin transform and some results which quarantee validity of formulas (A3.4), (A3.5).

Example A3.1.

a) $f(x)=e^{-x}$; $\hat{f}(s)=\Gamma(s)$ $(\sigma>0)$;

b) $f(x)=\begin{cases}1, & x<a\\ 0, & x\geq a\end{cases}$, $\hat{f}(s)=\frac{a^s}{s^2}$ $(\sigma>0)$;

c) $f(x)=\frac{1}{e^x-1}$; $\hat{f}(s)=\Gamma(s)\zeta(s)$, $(\sigma>1)$, where $\Gamma(s)$ is the Euler gamma-function, and $\zeta(s)$ is the Riemann zeta-function.

We will say $f(x)$ belongs to weighted space $L_2(\mathbb{R}_+,\rho)$, ρ is a positive measurable function, if

$$\int\limits_{0}^{+\infty}|f(x)|^2\rho(x)dx<+\infty.$$

Below we put $\rho(x)=x^{-1}$.

Proposition A3.5. *Let $x^k f(x)\in L_2(\mathbb{R}_+,x^{-1})$. Then*

$$\hat{f}(s,a)=\int\limits_{1/a}^{a}f(x)x^{s-1}dx \quad (\operatorname{Re}s=k)$$

converges in mean on straight line $(k - i\infty; k + i\infty)$ to some function $\hat{f}(s)$;

$$f(x, a) = \frac{1}{2\pi i} \int_{k-ia}^{k+ia} \hat{f}(s) x^{-s} ds$$

converges in mean to $f(x)$ in such sense that

$$\lim_{a \to \infty} \int_0^{+\infty} |f(x) - f(x, a)|^2 x^{2k-1} dx = 0,$$

and

$$\int_0^{+\infty} |f(x)|^2 x^{2k-1} dx = \frac{1}{2\pi} \int_{-\infty}^{+\infty} \left|\hat{f}(k + it)\right|^2 dt.$$

In other words $f(x, a) \to f(x)$ in norm of space $L_2\left(\mathbb{R}_+, x^{2k-1}\right)$. This result can be obtained from Plancherel's theorem with the help of ordinary change.

Proposition A3.6. *Let $x^k f(x)$, $x^{1-k} g(x)$ belong to $L_2(\mathbb{R}_+, x^{-1})$. Then*

$$\int_0^{+\infty} f(x) g(x) dx = \frac{1}{2\pi i} \int_{k-i\infty}^{k+i\infty} \hat{f}(s) \hat{g}(1 - s) ds.$$

Proposition A3.7. *Let $x^k f(x)$, $x^{\sigma-k} g(x)$ belong to $L_2(\mathbb{R}_+, x^{-1})$. Then*

$$f(x) g(x) \text{ and } \frac{1}{2\pi i} \int_{k-i\infty}^{k+i\infty} \hat{f}(\omega) \hat{g}(s - \omega) d\omega$$

is a pair of Mellin transforms in the sense that

$$\int_0^{+\infty} f(x) g(x) x^{s-1} dx = \frac{1}{2\pi i} \int_{k-i\infty}^{k+i\infty} \hat{f}(\omega) \hat{g}(s - \omega) d\omega$$

for all values t, $s = \sigma + it$.

The last result is obtained from proposition A3.6 by change $g(x)$ on $g(x)x^{s-1}$. The properties of the Mellin transform can be found for example in [66,249,251].

References

[1] Abdullaev, S.K., "Multidimensional singular integral equations in Hölder spaces with weight", *USSR Acad. Sci. Dokl. Math.*, **292**, 1987, 777-779 (Russian).

[2] Abdullaev, S.K., Vasil'ev, V.B., "On some classes of multidimensional singular integral equations over half-space in weighted Hölder scale", *Singular integral operators*, Baku, Azebijan St. Univ., 1986, 3 - 15 (Russian).

[3] Abdullaev, S.K., Vasil'ev, V.B., "Regularization of multidimensional singular integral equations on domain in Hölder spaces with weight," *Singular integral operators*, Baku, Azerbaijan St. Univ., 1989, 3 - 17 (Russian).

[4] Abdullaev, S.K., Vasil'ev, V.B., "On invertibility of multidimensional singular integral operators over half-space in weighted Hölder classes", *Differential Equations*, **26**, 1990, 1408 - 1416. (Russian)

[5] Agranovich, M.S., "Elliptic singular integro-differential operators", *Russian Math. Surveys*, 20:5, 1965, 3 - 120 (Russian).

[6] Agranovich, M.S., "Elliptic operators on compact manifolds", *Results in science and technics, Modern problems in mathematics, Fundamental branches*, Moscow, VINITI, **63**, 1990, 5 -129 (Russian).

[7] Alexandrov, V.M., Babeshko, V.A., "On pressure on elastic half-space by wedge shaped punch", *Appl. Maths Mechs*, **36**, 1972, 88 - 93 (Russian).

[8] Alexandrov, V.M., "Asymptotical methods in problems of continuum mechanics with mixed boundary conditions", *Appl. Maths Mechs*, **57**, 1993, 102 -108 (Russian).

[9] Alexandrov, V.M., Smetanin, B.I., Sobol, B.V., *Thin concentrators of stresses in elastic bodies*, Moscow, Nauka, 1993 (Russian).

[10] Atiyah, M.F., Singer, I.M., "Index of elliptic operators on compact manifolds", *Bull. Amer. Math. Soc.*, **69**, 1963, 422 - 433.

[11] Atiyah., M.F., *Lectures on K - theory*, Moscow, Mir, 1967 (Russian).

[12] Atiyah, M.F., "Algebraic topology and elliptic operators", *Comm Pure Appl. Math.*, **20**, 1967, 237 - 249.

[13] Atiyah, M.F., Singer, I.M., "Index of elliptic operators *I*," *Russian Math. Surveys*, 23:5, 1968, 99 - 142 (Russian).

[14] Atiyah, M.F., Segal, G.B., "Index of elliptic operators *II*," *Russian Math. Surveys*, 23:6, 1968, 135 - 149 (Russian).

[15] Atiyah, M.F., Singer, I.M., "Index of elliptic operators *III*," *Russian Math. Surveys*, 24:1, 1969, 127 - 182 (Russian).

[16] Atiyah, M.F., Singer, I.M., "Index of elliptic operators *IV*," *Russian Math. Surveys*, 27:4, 1972, 161 - 178 (Russian).

[17] Atiyah, M.F. Singer, I.M., "Index of elliptic operators *V*," *Russian Math. Surveys*, 27:4, 1972, 179 - 188 (Russian).

[18] Atiyah, M.F., "Global aspects of the theory of elliptic differential operators," *Proc. Int. Congr. Math. (Moscow - 1966)*, Moscow, Mir, 1968, 57 - 64.

[19] Atiyah M.F., "Algebraic topology and operators in Hilbert space," *Lect. Notes Math.*, **103**, 1969, 101 - 121.

[20] Atiyah, M.F., "Global theory of elliptic operators," *Proc. Intern. Conf. on Funct. Anal. and Relat. Topics, Tokyo*, 1969. Tokyo, 1970, 21 - 30.

[21] Atiyah, M.F., Bott R., "The index theorem for manifolds with boundary," *Bombay Coll. Diff. Analysis*, Oxford Univ. Press, 1964, 175 - 186.

[22] Akhiezer, A.I., Berestetskii, V.B., *Quantum electrodynamics*, Moscow, Nauka, 1981 (Russian).

[23] Babeshko, V.A., *Generalized factorization method in spatial dynamical mixed problems of elasticity theory*, Moscow, Nauka, 1984 (Russian).

[24] Babeshko, V.A., Glushkov, E.V., Zinchenko, Zh. F., *Dynamics of nonhomogeneous linear elastic medium*, Moscow, Nauka, 1989 (Russian).

[25] Belotserkovskii, S.M., Lifanov, I.K., *Numerical methods in singular integral equations and their application in aerodynamics, elasticity theory, electrodynamics*, Moscow, Nauka, 1985 (Russian).

[26] Berezin, F.A., *Method of secondary quantization*, Moscow, Nauka, 1986 (Russian).

[27] Besov, O.V., Il'in, V.P., Nikolskii, S.M., *Integral representations of functions and imbedding theorems*, Moscow, Nauka, 1975 (Russian).

[28] Bitsadze, A.V., *Boundary value problems for elliptic equations of second order*, Moscow, Nauka, 1966 (Russian).

[29] Bitsadze, A.V., *Some classes of partial differential equations*, Moscow, Nauka, 1981 (Russian).

[30] Bogol'ubov, N.N., Logunov, A.A., Oksak, A.I., Todorov, I.T., *General principles of quantum field theory*, Moscow, Nauka, 1987 (Russian).

[31] Bochner, S., Martin, U.T., *Functions of several complex variables*, Moscow, Publ. Foreign Lit., 1951 (Russian).

[32] Burchuladze, T.V., Gegelia, T.G., *Development of potential method in elasticity theory*, Tbilisi, Metsniereba, 1985 (Russian).

[33] Baxter, R.J., *Exactly solved models in statistical mechanics*, Academic Press, London - New York, 1982.

[34] Berger, C.A., Coburn, L.A., "Toeplitz operators and quantum mechanics," *J. Funct. Anal.*, **68**, 1986, 273 - 299.

[35] Boos, B., Bleecker, D.D., *Topology and Analysis. The Atiyah - Singer index formula and gauge - theoretic physics*, Springer - Verlag, New - York, Berlin, Heidelberg, Tokyo, 1985.

[36] Boos, B., Rempel, S., "Cutting and pasting of elliptic operators," *Math. Nachr.*, **109**, 1982, 157 - 194.

[37] Boos, B., Wojciechowski, K., "Desuspension of splitting elliptic symbols," *II*, Tekst IMFUFA, **113**, 1985, 1 - 31.

[38] Boos - Bavnbek, B., Wojciechowski, K.P., "Pseudo-differential projections and the topology of certain spaces of elliptic boundary value problems," *Commun. Math. Phys.*, **121**, 1989, 1 - 9.

[39] Böttcher, A., Silbermann, B., *Invertibility and asymptotics of Toeplitz matrices*, Berlin, Akademie - Verlag, 1983.

[40] Boutet de Monvel, L., "Boundary problems for pseudodifferential operators," *Acta Math.*, **126**, 1971, 11 - 51.

[41] Boutet de Monvel, L., Guillemin, V., *Spectral theory of Toeplitz operators*, Princeton, Princeton Univ. Press, 1981.

[42] Brüning, J., "Index theory for regular singular operators and applications," *Lect. Notes Math* ., **1256**, 1987, 36 - 54.

[43] Brüning, J., "On L_2 - index theorems for complete manifolds of rank-one type," *Duke Math. J.*, **66**, 1992, 257 - 309.

[44] Brüning, J., Seeley, R.T., "An index theorem for first order regular singular operators," *Amer. J. Math.*, **110**, 1988, 659 - 714.

[45] Calderon, A.P., "Uniqueness in the Cauchy problem for partial differential equations," *Amer. J. Math.*, **80**, 1958, 16 - 36.

[46] Calderon, A.P., "Cauchy integrals on Lipschitz curves and related operators," *Proc. Nat. Acad Sci. USA*, **74**, 1977, 1324 - 1327.

[47] Calderon, A.P., Zygmund,A., "On the existence of certain singular integrals," *Acta Math.*, **88**, 1952, 85 - 139.

[48] Calderon, A.P., Zygmund, A., "On singular integrals," *Amer. J. Math.*, **78**, 1956, 289 - 309.

[49] Calderon, A.P., Zygmund,A., "Singular integral operators and differential equations," *Amer. J. Math.*, **79**, 1957, 901 - 921.

[50] Calderon, A., "Boundary value problems for elliptic equations," *Soviet-American Symp. on Part. Differ. Equat. in Novosibirsk*, Moscow, Acad. Sci. USSR Press, 1963, 303 - 304 (Russian).

[51] Calderon, A.P., "Zygmund,A., On singular integrals with variable kernels," *Appl. Anal.*, **7**, 1978, 221 - 238.

[52] Cheeger, J., "Spectral geometry of singular Riemannian spaces," *J. Diff. Geom.*, **18**, 1983, 575 - 657.

[53] Coburn, L.A., Douglas, R.G., "C^* - algebras of operators on a half-space," *Publ. Math. IHES*, **40**, 1971, 59 -67.

[54] Coburn, L.A., Douglas, R.G., Schaeffer, D.G., Singer, I.M., "C^* - algebras of operators on a half-space. *II*. Index theory," *Publ. Math. IHES*, **40**, 1971, 69 - 80.

[55] Coburn, L.A., Douglas, R.G., Singer, I.M., "An index theorem for Wiener-Hopf operators on the discrete quarter plane," *J. Diff. Geom.*, **6**, 1972, 587 - 593.

[56] Cohen, J.W., Boxma, O.J., *Boundary value problems in queueing system analysis*, Amsterdam - New York - Oxford, North Holland, 1983.

[57] Colton D., Kress, R., *Integral equation methods in scattering theory*, John Wiley & Sons, New York-Chichester-Brisbane-Toronto- Singapore, 1983.

[58] Connes,A., "Feuilletages et algebres d'operateurs," *Lect. Notes Math.*, **842**, 1981, 139 - 155.

[59] Cordes, H.-O., "An algebra of singular integral operators with two symbol homomorphisms," *Bull. Amer. Math. Soc.*, **75**, 1969, 37 - 42.

[60] Cordes, H.-O., "Pseudodifferential operators on a half-line," *J. Math. Mech.*, **18**, 1969, 893 - 908.

[61] Cordes H.-O., Doong, S.H., "The Laplace comparison algebra of spaces with conical and cylindrical ends," *Lect. Notes Math.*, **1256**, 1987, 55 - 90.

[62] Cordes, H.-O., Herman, E.A., "Gel'fand theory of pseudodifferential operators," *Amer. J. Math.*, **90**, 1968, 681 -717.

[63] Dauge, M., "Higher order oblique derivative problems on polyhedral domains," *Commun. Part. Differ. Equat.*, **14**, 1989, 1193 - 1227.

[64] Dauge, M., Nicaise, S., "Oblique derivative and interface problems on polygonal domains and networks," *Commun. Part. Differ. Equat.*, **14**, 1989, 1147 - 1192.

[65] David, G., "Opérateurs intégraux singuliers sur certaines courbes du plan complexe," *Ann. Sci. Ecole Norm. Super.*, **17**, 1984, 157 - 189.

[66] Ditkin, V.A., Prudnikov, A.P., *Integral transformations and operational calculus*, Moscow, Fizmatgiz, 1961 (Russian).

[67] *Development of the theory of contact problems in USSR*, Ed. L.A. Galin, Moscow, Nauka, 1976 (Russian).

[68] Dezin, A.A., "Invariant differential operators and boundary value problems," *Proc. Steklov Math. Inst.*, **68**, 1962 (Russian).

[69] Dmitriev, V.I., Zakharov,E.V., *Integral equations in boundary problems of electrodynamics*, Moscow St. Univ. Press, Moscow, 1987 (Russian).

[70] Douglas, R.G., *C^* - algebra extensions and K - homology*, Princeton Univ. Press, 1980.

[71] Douglas, R.G., "Elliptic invariants and operator algebras: toroidal examples," *London Math. Soc. Lect. Notes Ser.*, **135**, 1988, 61 - 79.

[72] Douglas, R.G., Howe, R., "On C^* - algebra of Toeplitz operators in quarter-plane," *Trans. Amer. Math. Soc.*, **158**, 1971, 203 - 217.

[73] Duduchava, R.V., "Integral convolution equations with discontinuous presymbols, singular integral equations with fixed singularity and their applications to mechanical problems," *Proc. Tbilisi Math. Inst.*, **60**, 1979, 2 - 136 (Russian).

[74] Duduchava, R.V., "General singular integral equations and basic problems of plane elasticity theory," *Proc. Tbilisi Math. Inst.* **82**, 1986, 45 - 89 (Russian).

[75] Duduchava, R.V., "On multidimensional singular integral operators. I : The half-space case," *J. Oper. Theory*, **11**, 1984, 41 - 76.

[76] Duduchava, R.V., "On multidimensional singular integral operators. II : The case of compact manifolds," *J. Oper. Theory*, **11**, 1984, 199 - 214.

[77] Duduchava, R., "On algebras generated by convolutions and discontinuous functions," *Integr. Equat. and Oper. Theory*, **10**, 1987, 505 - 530.

[78] Duduchava, R., Schneider, R., "The algebra of non-classical singular integral operators on half-space," *Integr. Equat. and Oper. Theory*, **10**, 1987, 532 - 553.

[79] Dybin, V.B., *Correct problems for singular integral equations*, Rostov on Don St. Univ. Press, 1988 (Russian).

[80] Dynin, A.S., "Index of elliptic operator on compact manifold," *Russian Math. Surveys*, 21:5, 1966, 233 - 248 (Russian).

[81] Dynin, A.S., "Inversion problem for singular integral operators: C^* - approach," *Proc. Nat. Acad. Sci. USA*, **75**, 1978, 4668 - 4670.

[82] Dynin, A., "Multivariable Wiener-Hopf operators. *I*. Representations," *Integr. Equat. and Oper. Theory*, **9**, 1986, 537 - 556.

[83] Dynin, A., "Multivariable Wiener-Hopf operators. *II*. Spectral topology and solvability," *Integr. Equat. and Oper. Theory*, **10**, 1987, 554 - 576.

[84] Dyn'kin, E.M., "Methods of theory of singular integrals (Hilbert transform and Calderon-Zygmund theory)," *Results in science and technics, Modern Probl. Math., Fundamental branches*, **15**, Moscow, VINITI, 1986, 197 - 292. (Russian)

[85] Dzhuraev, A.D., *Method of singular integral equations*, Moscow, Nauka, 1987 (Russian).

[86] Egorov, Yu.V., *Linear differential equations of main type*, Moscow, Nauka, 1984 (Russian).

[87] Egorov, Yu.V., "Microlocal analysis," *Results in science and technics, Modern Probl. Math., Fundamental branches*, **33**, Moscow, VINITI, 1988, 5 - 156. (Russian)

[88] Egorov, Yu.V., Shubin, M.A., "Linear partial differential equations. Basics of classical theory," *Results of science and technics, Modern Probl. Math., Fundamental branches*, **30**, Moscow, VINITI, 1988, 1 - 262. (Russian)

[89] Egorov, Yu.V., Shubin, M.A., "Linear partial differential equations. Elements of modern theory," *Results in science and technics, Modern Probl. Math., Fundamental branches*, **31**, Moscow, VINITI, 1988, 5 - 125. (Russian)

[90] Egorov, Yu.V., Schulze, B.-W., *Pseudodifferential operators, singularities, applications*, Basel, Birkhäuser-Verlag, 1997.

[91] "Elliptische operatoren auf singulären und nicht kompakten mannigfaltigkeiten," *Tagungsber. Math. Forschungsinst. Oberwolfach*, **22**, 1987, 1 - 21.

[92] Eskin, G.I., *Boundary value problems for elliptic pseudodifferential equations*, AMS, Providence, RI, 1981.

[93] Eskin, G.I., *Boundary value problems for elliptic pseudodifferential equations*, Moscow, Nauka, 1973 (Russian).

[94] Eskin, G.I., "Conjugation problem for equations of main type with two independent variables," *Trans. Moscow Math. Soc.*, **21**, 1970, 245 – 292 (Russian).

[95] Eskin, G., "Boundary value problems for second order elliptic equations in domains with corners," *Proc. Symp. Pure Math.*, **43**, 1985, 105 – 131.

[96] Eskin, G., "Index formulas for elliptic boundary value problems in plane domains with corners," *Trans. Amer. Math. Soc.*, **314**, 1989., 283 –348.

[97] Fabes, E.B., "Layer potential methods for boundary value problems on Lipschitz domains," *Lect. Notes Math.*, **1344**, 1988, 55 – 80.

[98] Fedosov, B.V., "Analytical index formulas of elliptic operators," *Trans. Moscow Math. Soc.*, **30**, 1974, 159 – 241. (Russian)

[99] Fedosov, B.V., "Analytical index formula of elliptic boundary value problem. *I*," *Matem. sbornik*, **93**, 1974, 62 – 89 (Russian).

[100] Fedosov, B.V., "Analytical index formula of elliptic boundary value problem. *II*," *Matem. sbornik*, **95**, 1974, 525 – 550 (Russian).

[101] Fedosov, B.V., "Analytical index formula of elliptic boundary value problem. *III*," *Matem. sbornik*, **101**, 1976, 380 – 401 (Russian).

[102] Fedosov, B.V., "Index theorems," *Results in science and technics, Modern Probl. Math., Fundamental branches*, **65**, Moscow, VINITI, 1991, 165 – 268. (Russian)

[103] Fock, V.A., *Problems of diffraction and propagation of electromagnetic waves*, Moscow, Soviet Radio, 1970 (Russian).

[104] Freed, D., Uhlenbeck, K., *Instantons and four dimensional manifolds*, Moscow, Mir, 1988 (Russian).

[105] Gakhov, F.D., *Boundary problems*, Moscow, Nauka, 1977 (Russian).

[106] Galin, L.A., *Contact problems of elasticity and viscoelasticity theory*, Moscow, Nauka, 1980 (Russian).

[107] Gohberg, I., Krein M.G., "Basic statements on defect numbers, root numbers and indices of linear operators," *Russian Math. Surveys*, 12:2, 1957, 43 –118 (Russain).

[108] Gohberg, I., Krupnik, N.Ya., *Introduction to the theory of one- dimensional singular integral operators*, Kishinev, Shtiintsa, 1973 (Russian).

[109] Grisvard, P., *Elliptic problems in non-smooth domains, Monographs and Studies in Mathematics*, **21**, Boston, Pitman, 1985.

[110] Grubb, G., *Functional calculus of pseudo-differential boundary problems*, Birkhäuser, Boston, 1986.

[111] Grubb, G., "Pseudo-differential boundary problems in L_p spaces," *Commun. Part. Differ. Equat.*, **15**, 1990, 289 - 340.

[112] Gullemin, V., "Toeplitz operators in n dimensions," *Integr. Equat. and Oper. Theory*, **7**, 1984, 145 - 205.

[113] Helton, J.W., Howe, R.E., "Integral operators: traces, index and homology," *Lect. Notes Math.*, **345**, 1973, 141 - 209.

[114] Hirsh, M., *Differential topology*, Moscow, Mir, 1979 (Russian).

[115] Hirschman, T., "Functional analysis in cone and edge Sobolev spaces," *Ann. Global Anal. and Geom.*, **8**, 1990, 167 - 192.

[116] Hirzebruch, F., *Topological methods in algebraic geometry*, Springer-Verlag, 1978.

[117] Hirzebruch, F., "Elliptische differentialoperatoren auf mannifaltigkeiten," *Veröff. Arbritsgemein. Forsch. Land. Nordrhein-Westfalen, Natur - und Ing. Giesellschaftswiss*, **157**, 1965, 33 - 60.

[118] Hönl, H., Maue, A.W., Westpfahl, K., *Theorie der beugung*, Springer-Verlag, Berlin-Göttingen-Heidelberg, 1961.

[119] Hörmander, L., "Symbolic calculus and differential equations," *18 Scand. Congr. Math. Aarhus, 18 - 22 Aug., 1980, Proc.*, Boston e.a., 1981, 56 - 81.

[120] Hörmander, L., *Analysis of partial differential operators*, **1** – **4**, Springer-Verlag, 1983.

[121] Hu Sze-Tsen, *Homotopy theory*, Academic Press, New York, London 1959.

[122] Husemoller, D., *Fibre bundles*, Mc Graw-Hill Book Company, New York, 1966.

[123] Il'inskii, A.S., Kravtsov, V.V., Sveshnikov, A.G., *Mathematical models in electrodynamics*, Moscow, Higher School, 1991 (Russian).

[124] Ji, R., Xia, J., "On the classification of commutator ideals," *J. Funct. Anal.*, **78**, 1988, 208 -232.

[125] Journe, J.-L., *Calderon-Zygmund operators, pseudodifferential operators and the Cauchy integral of Calderon*, Springer-Verlag, 1983.

[126] Juhl, A., "Index formulas for oblique derivative problems," *Math. Nachr.*, **125**, 1986, 153 -174.

[127] Kakichev, V.A., "Boundary linear conjugation problems for holomorphic functions in bicylindrical domains," *Function theory, functional analysis and their applications*, **5**, Kharkov Univ. Press, Kharkov, 1967, 37 - 58 (Russian).

[128] Karapetyants, N.K., Samko, S.G., *Equations with involutive operators and their applications*, Rostov on Don Univ. Press, 1988 (Russian).

[129] Kasparov, G.G., "Operator K - theory and its applications," *Results in science and technics, Modern Probl. Math., Newest achievements*, **27**, Moscow, VINITI, 1985, 3 – 32 (Russian).

[130] Kokotov, A.Yu., "Representations of C^* – algebras of pseudodifferential operators in polyhedron," *Algebra and Analysis*, 3:3, 1991, 57 – 77 (Russian).

[131] Komech, A.I., "Elliptic boundary value problems for pseudodifferential operators on manifold with conical points," *Matem. sbornik*, **86**, 1971, 268 – 298 (Russian).

[132] Komech, A.I., "Elliptic boundary value problems on manifolds with piecewise smooth boundary," *Matem. sbornik*, **92**, 1973, 89 – 134 (Russian).

[133] Komech, A.I., "Linear partial differential equations with constant coefficients," *Results in science and technics, Modern Probl. Math., Fundamental branches*, **31**, Moscow, VINITI, 1988, 127 – 261 (Russian).

[134] Kondrat'ev, V.A., "Boundary value problems for elliptic equations in domains with conic or angular points," *Trans. Moscow Math. Soc.*, 1967, **16**, 209 – 292 (Russian).

[135] Kondrat'ev, V.A., Oleinik, O.A., "Boundary value problems for partial differential equations in non-smooth domains," *Russian Math. Surveys*, 38:2, 1983, 3 – 76 (Russian).

[136] Kozlov, V.A., Maz'ya, V.G., Rossmann, J., *Elliptic boundary value problems in domains with point singularities*, AMS, Mathematical Surveys and Monographs, **52**, 1997.

[137] Krein, M.G., "Integral equations on half-line with kernel depending on difference of arguments," *Russian Math. Surveys*, 13:5, 1958, 3 – 120. (Russian)

[138] Krein, S.G., *Linear equations in Banach space*, Moscow, Nauka, 1971 (Russian).

[139] Krupnik, N.Ya., *Banach algebras with symbol and singular integral operators*, Kishinev, Shtiintsa, 1984 (Russian).

[140] Kudryavtsev, L.D., Nikol'skii, S.M., "Spaces of differentiable functions of several variables and imbedding theorems," *Results in science and technics, Modern Probl. Math., Fundamental branches*, **26**, Moscow, VINITI, 1990, 5 – 155 (Russian).

[141] Kupradze, V.D., *Potential methods in elasticity theory*, Moscow, Fizmatgiz, 1963 (Russian).

[142] Kupradze, V.D., *Boundary problems in vibration theory and integral equations*, Moscow, Gostehizdat, 1950 (Russian).

[143] Kupradze, V.D., Gegelia, T.G., Basheleishvili, M.O., Burchuladze, T.G., *Three dimensional problems in mathematical elasticity and thermo-elasticity theory*, Moscow, Nauka, 1976 (Russian).

[144] Ladyzhenskaya, O.A., Uraltseva, N.N., *Linear and quasilinear equations of elliptic type*, Moscow, Nauka, 1964 (Russian).

[145] Levin, L., *Wave guides theory. Methods for solving of wave guide problems*, Moscow, Radio i Svyaz, 1981 (Russian).

[146] Lewis, J.E., Parenti, C., "Pseudodifferential operators of Mellin type," *Commun. Part. Differ. Equat.*, **16**, 1991, 1615 - 1664.

[147] Lifanov, I.K., *Singular integral equations and discrete vortices*, Zeist, VSP, 1996.

[148] Lifanov, I.K., Poltavskii, L.N., "Generalized Fourier operators and their application to justification of certain numerical methods in aerodynamics," *Matem. sbornik*, 183:5, 1992, 79 - 114 (Russian).

[149] Lions, J.L., Magenes, E., *Non-homogeneous boundary value problems and applications*, Berlin, Springer-Verlag, 1972.

[150] Lopatinskii, Ya.B., *Introduction into modern theory of partial differential equations*, Kiev, Naukova dumka, 1980 (Russian).

[151] Lopatinskii, Ya.B., *Theory of general boundary value problems. Selected papers*, Kiev, Naukova dumka, 1984 (Russian).

[152] Lurie, A.I., *Spatial problems of elasticity theory*, Moscow, Gostehteoretizdat, 1955 (Russian).

[153] Malgrange, B., "Opérateurs intégraux singuliers unicité du probléme de Cauchy," *Séminare Schwartz, Secrétariat mathématique, 4-e année, 1955/1960*, Paris, 1960, exposés 1 - 10.

[154] Malyshev, V.A., *Random walks. Wiener-Hopf equations in quarter plane. Galois automorphisms*, Preprint, Moscow St. Univ. Press, 1970 (Russian).

[155] Malyshev, V.A., "Wiener-Hopf equations in a quarter plane, discrete groups and automorphic functions," *Matem sbornik*, 84:4, 1971, 499 - 525 (Russian).

[156] Malyshev, V.A., "Wiener-Hopf equations and their appication in probability theory," *Results in science and technics, Probab. Theor. Math. Stat. Theor. Cybern.*, **13**, Moscow, VINITI, 1976, 5 - 35 (Russian).

[157] Maslov, V.P., "Non-standard characteristics in asymptotical problems," *Russian Math. Surveys*, 38:6, 1983, 3 - 36 (Russian).

[158] Maslov, V.P., *Asymptotical methods for solving of pseudodifferential equations*, Moscow, Nauka, 1987 (Russian).

[159] Maslov, V.P., *Asymptotical methods and perturbation theory*, Moscow, Nauka, 1988 (Russian).

[160] Maz'ya, V.G., *Sobolev spaces*, Leningrad St. Univ. Press, 1985 (Russian).

[161] Maz'ya, V.G., "Boundary integral equations," *Results in science and technics, Modern Probl Math., Fundamental branches*, **27**, Moscow, VINITI, 1988, 5 - 130 (Russian).

[162] Maz'ya, V.G., "Classes of domains, measures and capacities in the theory of spaces of differentiable functions," *Results in science and technics, Modern Probl. Math., Fundamental branches*, **26**, Moscow, VINITI, 1990, 159 - 228 (Russian).

[163] Maz'ya, V.G., Plamenevsky, B.A., "Elliptic boundary value problems on manifolds with singularities," *Probl. Math. Anal.*, **6**, Leningrad St. Univ. Press, 1977, 85 - 145 (Russian).

[164] Maz'ya, V.G., Shaposhnikova, T.O., *Multipliers in spaces of differentiable functions*, Leningrad St. Univ. Press, 1986 (Russian).

[165] Mazzeo, R., "Elliptic theory of differential edge operators," *Commun. Part. Differ. Equat.*, **16**, 1991, 1615 - 1664.

[166] Meister, E., "Some classes of integral and integro-differential equations of convolutional type," *Lect. Notes Math.*, **827**, 1980, 182 - 228.

[167] Meister, E., "Some solved and unsolved canonical problems in diffraction theory," *Lect. Notes Math.*, **1285**, 1987, 320 - 336.

[168] Miester, E., Speck, F.-O., "A contribution to the quarter-plane problem in diffraction theory," *J. Math. Anal. and Appl.*, **130**, 1988, 223 - 236.

[169] Melrose, R.B., "Calculus of conormal distributions on manifolds with corners," *Duke Math. J.*, 65:3, 1991, 51 - 61.

[170] Melroze, R.B., Piazza, P., "Analytic K - theory on manifolds with corners," *Adv. Math.*, **92**, 1992, 1 - 26.

[171] Mikhlin, S.G., *Multidimensional singular integrals and integral equations*, Moscow, Fizmatgiz, 1962 (Russian).

[172] Mikhlin, S.G., Prössdorf, S., *Singular integral operators*, Berlin, Akademie-Verlag, 1986.

[173] Miranda, C., *Equazioni alle derivate parziali di tipo ellittico*, Springer-Verlag, Berlin-Göttingen-Heidelberg, 1955.

[174] Mischenko, A.S., *Vector bundles and their applications*, Moscow, Nauka, 1984. (Russian)

[175] Mittra, R., Lee, S., *Analytical methods of wave guides theory*, Moscow, Mir, 1974 (Russian).

[176] Mizohata, S., *Theory of partial differential equations*, Moscow, Mir, 1974 (Russian).

[177] Morozov, N.F., *Selected two-dimensional problems of elasticity theory*, Leningrad St. Univ. Press, 1978 (Russian).

[178] Mossakovskii, V.I., Kachalovskaya, N.E., Golikova, S.S., *Contact problems of mathematical elasticity theory*, Kiev, Naukova dumka, 1985 (Russian).

[179] Moyer, R.D., "Computation of symbols on C^* – algebras of singular integral operators," *Bull Amer. Math. Soc.*, **77**, 1971, 615 – 620.

[180] Muskhelishvili, N.I, *Singular integral equations*, Moscow, Nauka, 1968 (Russian).

[181] Muskhelishvili, N.I., *Some basic problems of mathematical elasticity theory*, Moscow, Acad. Sci. USSR Press, 1954 (Russian).

[182] Nazarov, S.A., Plamenevsky, B.A., *Elliptic problems in domains with piecewise smooth boundaries*, Walter de Gruyter, Berlin - New York, 1994.

[183] Nazarov, S.A., Plamenevsky, B.A., "Elliptic problems with radiation conditions on edges of boundary," *Matem sbornik*, 183:10, 1992, 13 – 44 (Russian).

[184] Nicaise, S., "Differential equations in Hilbert spaces and applications to boundary value problems in nonsmooth domains," *J. Funct. Anal.*, **96**, 1991, 195 – 218.

[185] Noble, B., *Methods based on the Weiner-Hopf technique for the solution of partial differential equations*, AMS, Chelsea, 1988.

[186] Novikov, S.P., Sternin, B.Yu., "Traces of elliptic operators on submanifolds and K – theory," *Acad. Sci. USSR. Dokl. Math.* **170**, 1966, 1265 – 1268 (Russian).

[187] Novikov, S.P., Sternin, B.Yu., "Elliptic operators and submanifolds," *Acad. Sci. USSR. Dokl. Math.*, **171**, 1966, 525 – 528 (Russian).

[188] Osher, S.J., "Discrete potential theory and Toeplitz operators on the quarter plane. *I*," *Indiana Univ. Math. J.*, **24**, 1975, 887 – 896.

[189] Osher, S.J., "Initial boundary value problems for hyperbolic systems in regions with corners. *I*," *Trans. Amer. Math. Soc.*, **176**, 1973, 141 – 164.

[190] Palamodov, V.P., "Distributions and Fourier analysis," *Results in science and technics, Modern Probl. Math., Fundamental branches*, **72**, Moscow, VINITI, 1991, 5 - 134 (Russian).

[191] Palais, R.S., "Seminar on the Atiyah-Singer index theorem, with contributions by A. Borel, F.E. Browder and R. Solovay," *Ann. Math. Stud.*, **57**, Princeton Univ. Press, 1965.

[192] Parton, V.Z., Perlin, P.I., *Integral equations of elasticity theory*, Moscow, Nauka, 1977 (Russian).

[193] Parton, V.Z., Perlin, P.I., *Methods of mathematical elasticity theory*, Moscow, Nauka, 1981.

[194] Pasenchuk, A.E., *Abstract singular operators*, Novocherkassk Polytechn. Inst. Press, 1993 (Russian).

[195] Pilidi, V.S., "To question on the index of bisingular integral operators," *Math. Anal. and its Appl.*, **7**, Rostov on Don St. Univ. Press, 1975, 123 - 136 (Russian).

[196] Plamenevsky, B.A., *Algebras of pseudodifferential operators*, Moscow, Nauka, 1986 (Russian).

[197] Plamenevsky, B.A., Rosenblum, G.V., "On the index of pseudodifferential operators with isolated singularities in symbols," *Algebra and Analysis*, 2:5, 1990, 165 - 188 (Russian).

[198] Plamenevsky, B.A., Rosenblum, G.V., "Topological characteristics of algebra of singular integral operators on circumference with discontinuous symbols," *Algebra and Analysis*, 3:5, 1991, 155 - 167 (Russian).

[199] Plamenevsky, B.A., Rosenblum, G.V., "Pseudodifferential operators with discontinuous symbols: K - theory and index formulas," *Funct. Anal. and its Appl.*, 26:4, 1992, 45 - 56 (Russian).

[200] Plamenevsky, B.A., Senichkin, V.N., "Representations of algebra of pseudodifferential operators with piecewise smooth symbols," *Izvestiya Vuz. Math.*, **11**, 1989, 66 - 74 (Russian).

[201] Plamenevsky, B.A., Yudovin, M.E., "The first boundary value problem for convolution operator in cone," *Matem sbornik*, 77:2, 1968, 222 - 250 (Russian).

[202] Popov, G.Ya., *Contact problems for linear deformed base*, Kiev-Odessa, Visha Shkola, 1982 (Russian).

[203] Prössdorf,S., *Some classes of singular integral equations*, Berlin, Akademie-Verlag, 1974.

[204] Prössdorf, S., "Linear integral equations," *Results in science and technics, Modern Probl. Math., Fundamental branches*, **27**, Moscow, VINITI, 1988, 5 - 130 (Russian).

[205] Rabinovich, V.S., "Multidimensional Wiener-Hopf equation for cones," *Theory of functions, Functional analysis and their Appl.*, **5**, Kharkov Univ. Press, 1967, 37 - 58 (Russian).

[206] Rabinovich, V.S., "Pseudodifferential equations in unbounded domains with conic structure at infinity," *Matem. sbornik*, 89:1, 1969, 77 - 96 (Russian).

[207] Radkevich, E.V., "On operator sheaves of contact Stephan problem," *Math. Notices*, 47:2, 1990, 89 - 101 (Russian).

[208] Radkevich, E.V., "Problem of dynamical angle for the Gibbs-Thomson law," *Russian Acad. Sci. Dokl. Math.*, 323:5, 1992, 841 - 846.

[209] Rempel, S., Schulze, B.-W., *Index theory of elliptic boundary problems*, Berlin, Akademie-Verlag, 1982.

[210] Roe, J., "Partitioning non-compact manifolds and the dual Toeplitz problem," *London Math. Soc. Lect. Notes. Ser.*, **135**, 1988, 187 - 228.

[211] Rvachev, V.L., "On pressure of wedge shaped punch on elastic half-space," *Appl. Maths Mechs*, **23**, 1959, 169 - 171 (Russian).

[212] Rvachev, V.L., Protsenko, V.S., *Contact problems of elasticity theory for non-classical domains*, Kiev, Naukova dumka, 1977 (Russian).

[213] Samko, S.G., *Hypersingular integrals and their applications*, Rostov on Don St. Univ. Press, 1984 (Russian).

[214] Schulze, B.-W., "Mellin expansions of pseudo-differential operators and conormal asymptotics of solutions," *Lect. Notes Math.*, **1256**, 1987, 378 - 401.

[215] Schulze, B.-W., "Corner Mellin operators and reduction of orders with parameters," *Ann. Sc. Norm. Super. Pisa, Cl. Sci.*, **16**, 1989, 1 - 81.

[216] Schulze, B.-W., "Mellin representations of pseudo-differential operators on manifolds with corners," *Ann. Global Anal. and Geom.*, **8**, 1990, 261 - 297.

[217] Schulze, B.-W., *Pseudodifferential operators on manifolds with singularities*, North-Holland, Amsterdam, 1991.

[218] Schulze, B.-W., *Pseudo-differential boundary value problems, conical singularities, and asymptotics*, Berlin, Akademie-Verlag, 1994.

[219] Schwartz, A.S., "Elliptic operators in quantum field theory," *Results in science and technics, Modern Probl. Math.*, **17**, Moscow, VINITI, 1981, 113 - 173 (Russian).

[220] Schwartz, A.S., *Quantum field theory and topology*, Moscow, Nauka, 1989 (Russian).

[221] Schwartz, J.T., *Differential geometry and topology*, Notes on Math. and its Appl., New York, 1968.

[222] Schwartz, L., *Complex manifolds. Elliptic equations*, Moscow, Mir, 1964 (Russian).

[223] Seeley, R.T., "Singular integrals on compact manifolds," *Amer. J. Math.*, **81**, 1959, 658 - 690.

[224] Seeley, R.T., "Regularisation of singular integral operators on compact manifolds," *Amer. J. Math.*, **83**, 1961, 265 - 275.

[225] Seeley, R.T., "Integro-differential operators on vector bundes," *Trans. Amer. Math. Soc.*, 117:5, 1965, 167 -204.

[226] Seeley, R.T., "Singular integrals and boundary value problems," *Amer. J. Math.*, **88**, 1966, 781 - 809.

[227] Senichkin, V.N., "On Jacobson topology on spectra of algebra of Wiener-Hopf operators in cone," *Algebra and Analysis*, 5:6, 1993, 194 - 218 (Russian).

[228] Shtaerman, I.Ya., *Contact problem of elasticity theory*, Moscow-Leningrad, Gostehteoretizdat, 1949 (Russian).

[229] Shubin, M.A., *Pseudodifferential operators and spectral theory*, Moscow, Nauka, 1978 (Russian).

[230] Simonenko, I.B., "New general method for investigation of linear operator equations of singular integral equations type. *I*," *Izvestiya Acad. Sci. USSR, Ser. math.*, **29**, 1965, 567 - 586; *II*, ibidem, 757 - 782 (Russian).

[231] Simonenko, I.B., "Convolution type operators in cones," *Matem. sbornik*, **74**, 1967, 298 - 313 (Russian).

[232] Singer, I.M., "Future extensions of index theory and elliptic operators," *Ann. Math. Stud.*, **70**, 1971, 171 - 185.

[233] Singer, I.M., "Some remarks on operator theory and index theory," *Lect. Notes Math.*, **575**, 1977, 128 - 138.

[234] Sobolev, S.L., *Some applications of functional analysis in mathematical physics*, Moscow, Nauka, 1988 (Russian).

[235] Sobolev, S.L., *Selected questions of functional spaces and distribution theory*, Moscow, Nauka, 1989 (Russian).

[236] Soldatov, A.P., *One-dimensional singular operators and boundary problems of functions theory*, Moscow, Higher School, 1991 (Russian).

[237] Soldatov, A.P., *Boundary value problems of functions theory in domains with piecewise smooth boundary*, Parts 1,2, Tbilisi Univ. Press, 1991 (Russian).

[238] Solonnikov, V.A., Frolova, E.V., "Oblique derivative problem for the Laplace equation in plane angle," *Nonlinear boundary value problems*, **3**, Kiev, Naukova dumka, 1991, 86 - 94 (Russian).

[239] Solonnikov, V.A., Frolova, E.V., "Studying of problem for the Laplace equation with boundary condition of special type in a plane angle," *Notes Sci. Semin. Leningrad branch of Steklov Math. Inst.*, **182**, 1990, 149 - 167 (Russian).

[240] Speck, F.-O., *General Wiener-Hopf factorization methods*, Pitman Advanced Publishing Program, Boston-London-Melbourne, 1985.

[241] Stein, E.M., *Singular integrals and differentiability properties of functions*, Princeton Univ. Press, Princeton, 1970.

[242] Stein, E.M., "Some problems of harmonic analysis related to curvature and osillator integrals," *Int. Congr. Math., Berkley, Calif., 1986, Survey reports*, Moscow, Mir, 1991, 297 - 332 (Russian).

[243] Stein, E.M., Weiss, G., *Introduction to Fourier analysis on Euclidean spaces*, Princeton Univ. Press, Princeton, 1971.

[244] Sternin, B.Yu., *Topological aspects of S.L. Sobolev problem*, Moscow Inst. Electronic Mach., 1971 (Russian).

[245] Sternin, B. Yu., *Elliptic theory on compact manifolds with singularities*, Moscow, Moscow Inst. Electronic Mach., 1974 (Russian).

[246] Taylor, M., *Pseudodifferential operators*, Princeton Univ. Press, Princeton, 1981.

[247] Teleman, M., "The index theorem for topological manifolds," *Acta Math.*, **153**, 1984, 117 - 152.

[248] Tikhonov, A.N., Samarskii, A.A., *Equations of mathematical physics*, Moscow, Nauka, 1972 (Russian).

[249] Titchmarsch, E., *Introduction to Fourier integrals theory*, Moscow-Leningrad, Gostehteoretizdat, 1948 (Russian).

[250] Treves, F., *Introduction to pseudodifferential and Fourier integral operators*, **1,2**, Plenum Press, New York-London, 1982.

[251] Uflyand, Ya.S., *Integral transformations in problems of elasticity theory*, Leningrad, Nauka, 1967 (Russian).

[252] Uflyand, Ya.S., *Method of paired equations in problems of mathematical physics*, Leningrad, Nauka, 1977 (Russian).

[253] Upmeier, H., "An index theorem for multivariable Toeplitz operators," *Integr. Equat. and Oper. Theory*, **9**, 1986, 355 - 586.

[254] Upmeier, H., "Index theory for multivariable Weiner-Hopf operators," *J. reine und angew. Math.*, **384**, 1988, 57 -79.

[255] Upmeier, H., "Fredholm indices for Toeplitz operators on bounded symmetric domains," *Amer. J. Math.*, **110**, 1988, 811 - 832.

[256] Vainshtein, L.A., *Electromagnetic waves*, Moscow, Radio i Svyaz, 1986 (Russian).

[257] Vainshtein, L.A., *Theory of diffraction and factorization method*, Moscow, Soviet Radio, 1966 (Russian).

[258] Vasilevskii, N.L., "Two-dimensional operators of Mikhlin-Calderon-Zygmund and bisingular operators," *Siberian Math. J.*, 27:2, 1986, 23 - 31 (Russian).

[259] Vasilevskii, N.L., "Algebras generated by multi-dimensional singular integral operators and coefficients admitting discontinuity of homogeneous type," *Matem. sbornik*, 129:1, 1986, 3 - 19 (Russian).

[260] Vasilevskii, N.L., "On algebra related to Toeplitz operators in radial tube domains," *Izvestiya Acad. Sci. USSR, Ser. math.*, **51**, 1987, 79 - 95 (Russian).

[261] Vasilevskii, N.L., "Hardy spaces related to Siegel domains," *Repts enlarged Semin. I.N. Vekua Inst. Appl. Math.*, 3:1, 1988, 48 - 51 (Russian).

[262] Vasil'ev, V.B., "Singular integrals on compact manifolds with singularities," *Differential Equations*, **29**, 1993, 1427 - 1428.

[263] Vasil'ev, V.B., "Singular integral operators in a domain with corners on a plane," *Differential Equations*, **30**, 1994, 1426 - 1435.

[264] Vasil'ev, V.B., "Many-dimensional Riemann problem and related singular integral equations," *Differential Equations*, **31**, 1995, 492 - 494.

[265] Vasil'yev, V.B., "The Wiener-Hopf equations and the mathematical theory of diffraction," *Comp. Maths Math. Phys.*, **34**, 1994, 1635 - 1641.

[266] Vasil'yev, V.B., "An integral equation for a problem of the indentation of a wedge-shaped punch," *J. Appl. Maths Mechs*, **59**, 1995, 255 - 261.

[267] Vasil'ev, V.B., "Pseudodifferential equations in cones," *Differential Equations*, **31**, 1995, 1335 - 1346.

[268] Vasil'ev, V.B., "Boundary problems for some classes of pseudodifferential equations in a plane infinite angle," *Proc. the 6th Symp. Math. and its Appl.*, Techn. Univ of Timişoara, 2 - 5 Nov. 1995, Timişoara, Romania, 190 - 195.

[269] Vasil'ev, V.B., "Integral equations related to boundary value problems for the Laplace equation in a plane sector," *Differential Equations*, **32**, 1996, 1187 - 1193.

[270] Vasil'ev, V.B., "On general boundary value problems for some classes of pseudodifferential equations in a quarter-plane," *Differential Equations*, **33**, 1997, 995 - 997 (Russian).

[271] Vasil'ev, V.B., "On Dirichlet and Neumann problems for some classes of pseudodifferential equations in a quarter-plane," *Differential Equation*, **33**, 1997, 1136 - 1138 (Russian).

[272] Vasil'ev, V.B., "Potential type integral operators related to elliptic problems in non-smooth domains," *Differential Equations*, **33**, 1997, 1167 - 1173 (Russian).

[273] Vasil'ev, V.B., "On some classes of pseudodifferential equations in cones," *Russian Acad. Sci. Dokl. Math.*, 355:1, 1997, 14 - 17 (Russian).

[274] Vasil'ev, V.B., "Singular integral operators, wave factorization and boundary value problems," *Actual Probl. Modern Math.*, **3**, Novosibirsk St. Univ. Press, 1997, 39 - 45 (Russian).

[275] Vasil'ev, V.B., "Regularization of multidimensional singular integral equations in non-smooth domains," *Trans. Moscow Math. Soc.*, **59**, 1998, 65 - 93.

[276] Vasil'ev, V.B., "Boundary value problems and integral equations for some classes of pseudodifferential equations in non-smooth domains," *Differential Equations*, 34:9, 1998 (Russian).

[277] Vasil'ev, V.B., "On "normal" solvability of general boundary value problems for some classes of pseudodifferential equations," *Differential Equations*, 1999, **35** (in press).

[278] Vasil'ev, V.B., "Wave factorization of elliptic symbols," *Math. Notices* (in press).

[279] Vasil'ev, V.B., *Boundary value problems for the Laplacian in infinite plane angle*, Siberian Math. J. (in press).

[280] Verchota, G., "Layer potential and regularity for the Dirichlet problem for Laplace's equation in Lipschitz domains," *J. Funct. Anal.*, **59**, 1984, 572 - 611.

[281] Vishik, M.I., "Elliptic convolution equations in a bounded domain and their applications," *Proc. Int. Congr. Math., Moscow, 1966*, Moscow, Mir, 1968, 409 - 420 (Russian).

[282] Vishik, M.I., Eskin, G.I., "Convolution equations in a bounded domain," *Russian Math. Surveys*, 20:3, 1965, 89 - 152 (Russian).

[283] Vladimirov, V.S., "Linear conjugation problem for holomorphic functions," *Izvestiya Acad. Sci. USSR, Ser. math.*, **29**, 1965, 807 – 834 (Russian).

[284] Vladimirov, V.S., "Linear conjugation problem for holomorphic functions of many complex variables," *Modern Problems of Analytical Functions Theory*, Moscow, Nauka, 1966, 64 – 68 (Russian).

[285] Vladimirov, V.S., *Methods of functions theory of many complex variables*, Moscow, Nauka, 1964 (Russian).

[286] Vladimirov, V.S., *Distributions in mathematical physics*, Moscow, Nauka, 1979 (Russian).

[287] Vladimirov, V.S., Drozhzhinov, Yu.N., Zav'yalov, B.I., *Multidimensional Tauberian theorems for distributions*, Moscow, Nauka, 1986 (Russian).

[288] Vladimirov, V.S., Sergeev, A.G., "Complex analysis in future tube," *Results in science and technics, Modern Probl. Math., Fundamental branches*, V. 8, Moscow, VINITI, 1985, 191 – 266. (Russian).

[289] Vol'pert, A.I., "On index and normal solvability of boundary value problems for elliptic systems of differential equations on a plane," *Trans. Moscow Math. Soc.*, **10**, 1961, 41 – 87 (Russian).

[290] Vol'pert, A.I., "Elliptic systems on sphere and two-dimensional singular equations," *Matem sbornik*, **59** (suppl.), 1962, 195 – 214 (Russian).

[291] Vorovich, I.I., Babeshko, V.A., *Dynamical mixed problems of elasticity theory for non-classical domains*, Moscow, Nauka, 1979 (Russian).

[292] Vorovich, I.I., Alexandrov, V.M., Babeshko, V.A., *Non-classical mixed problems of elasticity theory*, Moscow, Nauka, 1974 (Russian).

[293] "Wiener-Hopf problems with applications," *Tagungsber. Math. Forschungsinst.*, **31**, Oberwolfach, 1986, 1 – 15.

[294] Wojciechowski, K., "Elliptic operators and relative K – homology groups on manifolds with boundary," *Repts Acad. Sci. Can.*, 7:2, 1985, 149 – 154.

[295] Wojciechowski, K., "Spectral flow and the general linear conjugation problem," *Simon Stevin*, **59**, 1985, 59 – 91.

[296] Wojciechowski, K.P., "On the Calderon projections and spectral projections of the elliptic operators," *J. Oper. Theory*, **20**, 1988, 107 – 115.

[297] Yanushauskas, A., *Potential methods in the theory of elliptic equations*, Vilnius, Mokslas, 1990 (Russian).

Index